T0250808

PoliTO Springer Series

Editor-in-Chief

Giovanni Ghione, Dept. of Electronics and Telecommunications, Politecnico di Torino, Italy

Editorial Board

Andrea Acquaviva, Dept. of Control and Computer Engineering, Politecnico di Torino, Italy
Pietro Asinari, Dept. of Energy, Politecnico di Torino, Italy
Claudio Canuto, Dept. of Mathematical Sciences, Politecnico di Torino, Italy
Erasmo Carrera, Dept. of Mechanical and Aerospace Engineering, Politecnico di Torino, Italy
Felice Iazzi, Dept. of Applied Science and Technology, Politecnico di Torino, Italy
Luca Ridolfi, Dept. of Environment, Land and Infrastructure Engineering, Politecnico di Torino, Italy

Springer, in cooperation with Politecnico di Torino, publishes the PoliTO Springer Series. This co-branded series of publications includes works by authors and volume editors mainly affiliated with Politecnico di Torino and covers academic and professional topics in the following areas: Mathematics and Statistics, Chemistry and Physical Sciences, Computer Science, All fields of Engineering. Interdisciplinary contributions combining the above areas are also welcome. The series will consist of lecture notes, research monographs, and briefs. Lectures notes are meant to provide quick information on research advances and may be based e.g. on summer schools or intensive courses on topics of current research, while SpringerBriefs are intended as concise summaries of cutting-edge research and its practical applications. The PoliTO Springer Series will promote international authorship, and addresses a global readership of scholars, students, researchers, professionals and policymakers.

More information about this series at http://www.springer.com/series/13890

Angelo Angelini

Photon Management Assisted by Surface Waves on Photonic Crystals

POLITECNICO
DI TORINO

Angelo Angelini
DISAT
Politecnico di Torino
Turin
Italy

ISSN 2509-6796   ISSN 2509-7024   (electronic)
PoliTO Springer Series
ISBN 978-3-319-84325-4   ISBN 978-3-319-50134-5   (eBook)
DOI 10.1007/978-3-319-50134-5

© Springer International Publishing AG 2017
Softcover reprint of the hardcover 1st edition 2017
This work is subject to copyright. All rights are reserved by the Publisher, whether the whole or part
of the material is concerned, specifically the rights of translation, reprinting, reuse of illustrations,
recitation, broadcasting, reproduction on microfilms or in any other physical way, and transmission
or information storage and retrieval, electronic adaptation, computer software, or by similar or
dissimilar methodology now known or hereafter developed.
The use of general descriptive names, registered names, trademarks, service marks, etc. in this
publication does not imply, even in the absence of a specific statement, that such names are exempt
from the relevant protective laws and regulations and therefore free for general use.
The publisher, the authors and the editors are safe to assume that the advice and information in this
book are believed to be true and accurate at the date of publication. Neither the publisher nor the
authors or the editors give a warranty, express or implied, with respect to the material contained
herein or for any errors or omissions that may have been made.

Printed on acid-free paper

This Springer imprint is published by Springer Nature
The registered company is Springer International Publishing AG
The registered company address is: Gewerbestrasse 11, 6330 Cham, Switzerland

A Deliana, compagna di una vita.

Acknowledgements

This book is the summary of 3 years of work and during this period I met several people that somehow influenced the outcome of my Ph.D. activity, directly or indirectly. Nevertheless, I would like to mention here some of the people who have been fundamental for the results I achieved.

A special thanks goes to Prof. Emiliano Descrovi, who has believed in me since the beginning and has always pushed me to do my best. Basically he taught me everything I know about optics and photonics, and our discussions have always been inspirational for me. It is quite fair to say that this work is also his work.

Dr. Natascia De Leo, Dr. Luca Boarino, Dr. Emanuele Enrico and all the people at INRIM for their constant essential technological support.

My mother, for her discrete and constant presence. I know you can understand my mood even 1000 km far away, without any need for many words. Thanks.

My father, who has been providing me everything I needed along all these years.

At last, I would like to thank her who has always believed in me, keeping in spurring me to do always my best. This book is dedicated to you.

Contents

List of Figures

Introduction

In the recent past, the emerging field of nanotechnologies has stimulated intense research efforts, since it holds promise for opening new scenarios in a wide variety of fields. Although *"this field [...] will not tell us much more about fundamental physics,"* as stated by Feynman in 1959, *"it might tell us much more about all the strange phenomena that occur in complex situations. Furthermore [...], it would have an enormous number of technical applications."* As expected by Feynman, fifty years later, nanotechnologies have become pervasive in almost all the research fields, from biology to medicine, from materials science to information technologies just to give few examples.

Within the world of nanotechnologies, nanophotonics and, more generally, technologies devoted to the manipulation of light have gained relevance in several fields. The conversion of far-field electromagnetic radiation into localized energy and vice versa, as well as the control of the radiation angular pattern of energy emitted by localized sources, is of outstanding relevance in a wide variety of fields.

In the radio frequency regime, for example, a plethora of applications make use of antennas to transmit and receive information, from satellite communications to mobile phones. The extension of such approach to the visible range of frequencies has led to develop the novel concept of optical antennas.

A general argument in antennas theory is the scalability of parameters; that is, the antenna parameters are determined by the wavelength of incident radiation. For a long time, the main challenge in designing antennas for visible light has been related to fabrication capabilities, since the typical size of an optical antenna should be of the order of tens of nanometers, with a resolution of few nanometers.

Thanks to the development, during the last decades, of nanofabrication techniques such as electron-beam lithography (EBL) or focused ion beam (FIB) lithography, a wide variety of optical antennas have been proposed in the recent past, based on dielectric and metallic nanostructures. Particular attention have gained the metallic ones, since the exploitation of surface plasmon resonances (SPRs) allows for confining the electromagnetic field in deep subwavelength volumes and allows for controlling the radiation patter, making such structures effective optical antennas. Moreover, SPRs are localized at the interface between a metal and a dielectric medium, making them suitable as transducers for environment variations and allowing an easy integration of photon sources with the antenna itself. Unfortunately, SPR relies on the oscillations of free electrons in metal, and it is therefore intrinsically affected by the scattering of electrons with metallic ions. The scattering results in ohmic losses that lower the performances of SPR-based devices.

Another strategy widely explored to control the light propagation is the exploitation of dielectric materials properly designed, also known as photonic crystals (PCs). Photonic crystals are periodic dielectric structures whose periodicity is comparable with the electromagnetic wavelength. Due to low intrinsic losses in dielectric materials at the visible wavelengths, photonic crystal resonators show narrow resonances with quality factors that can be orders of magnitude higher compared to metallic resonators. For this reason, PCs are widely used as high-quality cavity resonator, reflectors, or waveguides. Although the propagation of light can be strongly manipulated and finely controlled by means of PCs, the request for deep subwavelength confinement cannot be satisfied by purely dielectric structures, because of the diffraction limit. Moreover, the electromagnetic field is usually buried inside the photonic structures, making it difficult the interaction of the environment with the photonic mode. In the attempt to overcome the limitations of the two approaches, several hybrid metallo-dielectric structures have been proposed, with the aim of integrating the confining and enhancing properties of plasmonic structures with the high quality factors of photonic ones.

In this work, I will discuss the use of purely dielectric structures as a valuable alternative to metallic thin films sustaining SPRs. Properly designed one-dimensional photonic crystals (1DPCs) can indeed sustain electromagnetic modes localized at the truncation interface of periodic stacks of dielectric thin films. Their phenomenological behavior is similar to surface plasmon polaritons (SPPs) except from the lack of intrinsic ohmic losses which results in higher performances of the resulting device. Other inherent advantages such as the wide tunability in terms of resonant wavelengths and polarization state of the mode will be discussed as well.

Chapter 1 deals with a description of the basic principles underlying the physics of BSWs on 1DPC and with the interaction between BSW and surface structures. In Chap. 2, the interaction of spontaneous emitters with the photonic environment is discussed. Chapter 3 deals with a particular surface structure, i.e. a ring antenna, which is capable of focusing light, and redirects luminescence coming from localized sources into a collimated beam.

Chapter 1
Bloch Surface Waves on A One Dimensional Photonic Crystal

The first observations of Lord Rayleigh about the reflective properties of certain crystals of chlorate of potash in 1888 [1], led him to hypothesize that "*on the whole, the character of the reflected light appears to me to harmonize generally with the periodical theory*". In that paper, he commented on the peculiar internal color observed in the crystals, arguing that the phenomenon could be attributed to an internal periodic structure acting as a grating. Coherent superposition causing constructive interference along certain directions for certain wavelengths was responsible for the observed coloration. The wave-like nature of light is indeed fully coherent with the observation of interference effects in periodic structures or in thin film. Unfortunately, at that time Lord Rayleigh could only hypothesize the effect of periodicity on the photonic properties of crystals. Nowadays, thanks to microscopic imaging techniques, it is known that also many biological systems exploit a sub-micrometric structuration of dielectric materials in order to take advantage of their photonic properties [2, 3]. Iridescence of butterfly wings or beetle shells are few examples.

Some decades later, the theory of quantum mechanics, by means of the Bloch Theorem, successfully explained the propagation of electrons in a crystal lattice in terms of traveling waves in a periodic potential. It was demonstrated that a periodic ionic potential gives rise to an electronic band structure, i.e. a set of states with well defined energy and momentum. Electrons, described by probability wavefunctions, occupie states according to the dispersion relation originated by the periodic potential. The same picture could be applied to the propagation of photons in photonic crystals [4], provided that electrons (fermions) and photons (bosons) follow different statistics, namely the Fermi-Dirac distribution and the Bose-Einstein distribution respectively. If the ionic potential, that in solids gives rise to the electronic band structure is replaced by the dielectric function and the electronic wave-function is replaced by an electromagnetic wave, it can be shown that a periodic dielectric structure produces a photonic band structure [4]. The periodicity of the refractive index may concern one, two or three directions, and the systems associated are consequently called one, two or three dimensional photonic crystals (Fig. 1.1). The dimensionality of the photonic band structure reflects the dimensionality of the periodicity of the dielectric function.

© Springer International Publishing AG 2017
A. Angelini, *Photon Management Assisted by Surface Waves on Photonic Crystals*, PoliTO Springer Series, DOI 10.1007/978-3-319-50134-5_1

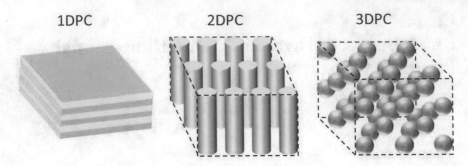

Fig. 1.1 Schematic view of photonic crystals with different dimensionality

Let's consider the one dimensional periodic structure constituted by 6 pairs of layers with alternating high refractive index ($n_h = 2$) and low refractive index ($n_l = 1.5$), each of them 100 nm thick, shown in Fig. 1.2. We can define the transverse wave-vector:

$$k_T = \sqrt{k_x^2 + k_y^2} \tag{1.1}$$

where k_x and k_y are the wavevector components along x and y respectively.

The transverse wave-vector k_T accounts for the angle of incidence of a plane wave with respect to the z-axis, and can be expressed as:

$$k_T = k_0 n \sin\theta_{inc} \tag{1.2}$$

where k_0 is the free-space wave-vector and n is the refractive index of the external homogeneos dielectric medium.

In general the response of the photonic structure depends on the polarization of the incident beam. Let's define the plane of incidence as the plane containing both the direction of propagation of the beam (k_{inc}) and the normal to the surface ($\hat{s}$) (see Fig. 1.2). It is now possible to define two orthogonal polarization states that will constitute a basis for all the possible polarization states: the transverse magnetic (TM) polarization (or equivalently the p-polarization) corresponds to an electric field oscillating within the incidence plane (red arrow), while the Transverse Electric (TE) polarization (also called s-polarization) corresponds to an electric field oscillating out of the incidence plane (green arrow).

For each polarization state we can now build up reflectance or transmission maps as function of the wavelength (analogous to energy) and angle of incidence. As an example the reflectance maps for TE and TM polarization of the 1DPC schematized in (Fig. 1.2) are shown in Fig. 1.3.

The two reflectance maps show a region of high reflectance centered at 700 nm at normal incidence (note that in this specific case 700 nm is exactly half the optical path of the elementary cell, as expected from the Bragg's law). The width of the stop band can be tuned by tailoring the refractive index contrast (the higher the contrast,

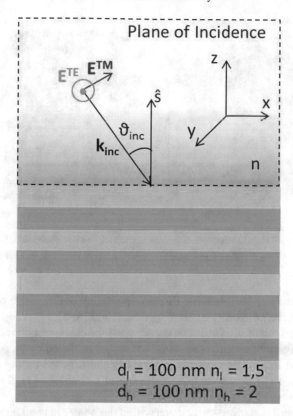

Fig. 1.2 In the *upper* homogenous region, a scheme that defines the plane of incidence when an electromagnetic plane wave impinges on a surface. The surface is characterized by its normal vector $\hat{s}$. The plane wave is characterized by a wave-vector (k_{inc}) and the electric field, whose orientation can be decomposed in two orthogonal states: TM (*red arrow*) and TE (*green point*). In the *lower* region: sketch of an exemplary multilayer stack composed by high (n = 2) and low (n = 1.5) refractive index layers each 100 nm thick

the larger the stop band), while the maximum reflectivity is given by the number of elementary cells. If the periodicity is not limited to a single dimension, the stop band may concern two or three dimensions. In the latter case, a complete stop band can be opened, meaning that the propagation of light of a specific range of frequencies is inhibited along all possible directions.

Many of the most remarkable effects in crystalline structures are related with the formation of forbidden gaps within the band structure (Fig. 1.2). The high reflectivity along certain directions of different wavelengths, as the one observed by Rayleigh in the crystals of chlorate of potash, can be explained by looking at the dispersion diagram of the crystals. Structuring the dielectric function at the optical wavelengths scale allows for a certain degree of manipulation of light, as well as the control over the crystalline structures of solids allows for controlling their electronic properties. For example, it has been shown in literature that the fabrication of photonic structures

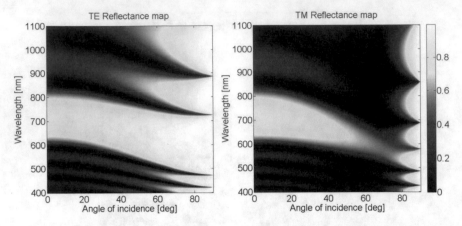

Fig. 1.3 Exemplary TE and TM reflectance maps associated to a one dimensional photonic crystal. The maps show a region of high reflection corresponding to the band gap of the photonic crystal

allows for controlling the photonic dispersion relation and thus the propagation of light, including the possibility of slowing down light [5, 6].

The increased capability, in the last decades, to fabricate structures at the nanoscale and the high degree of control of material properties achieved by modern deposition systems has led to a plethora of photonic applications based on the control of the dispersion relation of light in solids, such as laser technology [7], lighting systems [8] and many others [9, 10].

Trapping states, i.e. evanescent states within the photonic band gap, can be eventually created by inserting defects in the crystalline structure [11]. In this case it is possible to confine photons within an optical cavity by properly tailoring the defect dimensions [12]. Photon confinement means that the energy density function shows the maximum of intensity inside the cavity and an exponentially decaying profile in all directions. When properly tailored, confined states can resonate at specific wavelengths with extremely high Q-factors, increasing significantly the Local Density of Optical States (LDOS) [13]. Such effect can be also exploited to control spontaneous emission decay rates [14–16].

The radiative decay of an emitter can be modeled as a dipolar transition from an initial state Ψ_i to the final state Ψ_f through the emission of a photon. The transition probability is governed by the Fermi Golden rule:

$$\gamma_R(\vec{r}, \omega, \vec{d}) = \frac{\pi \omega}{\hbar \epsilon_0} \mid \langle \Psi_f \mid d \mid \Psi_i \rangle \mid^2 \rho(\vec{r}, \omega, \vec{d}) \tag{1.3}$$

where γ_R is the radiative decay rate, $\vec{r}$ is the spatial position of the emitter, ω is the angular frequency of the photon emitted, $\vec{d}$ is the dipole momentum and ρ is the LDOS at a given position in the space, for a given energy and for a given orientation of the dipole.

The Fermi Golden Rule states that the photonic environment influences the transition probability by modifing the distribution of available photonic states: an inhomogeneity of the LDOS in the wave-vector space will influence the angular radiation pattern, whilst a variation in the number of available photonic states modifies the overall transition probability, thus increasing or reducing the radiative decay rate. The influence of the photonic environment on the behaviour of spontaneous emitters, also known as the Purcell effect [17], has been demonstrated and widely exploited in the so-called optical antennas [18]. A complete suppression of the spontaneous emission has been also demonstrated in the case of a 3D photonic crystal showing a complete band gap [16, 19].

1.1 Bloch Surface Waves: General Properties

In the following I will focus my attention on 1DPCs, also known as Distributed Bragg Reflectors (DBRs) or Bragg mirrors. Such systems always exhibit a stop band, or forbidden band, in a certain energy-momentum range.

Let's consider now a 1DPC grown on a glass substrate (refractive index $n_{sub} = 1.5$) terminated by a semi-infinite homogenous dielectric layer (we assume the superstrate is air, so that $n_{sup} = 1$). When light impinges on the 1DPC from the glass side, the reflectance map will show a region of complete reflection indipendent on the plane wave wavelength (Fig. 1.4).

Fig. 1.4 TE reflectance map of the ML sketched in Fig. 1.2 grown on a glass substrate. The map clearly shows the region of Total Internal Reflection beyond the critical angle

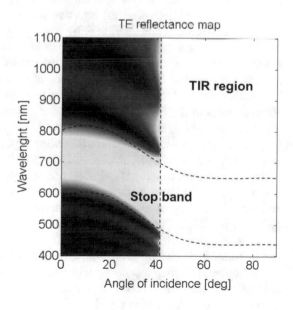

The condition which sets the limit angle (also referred as the critical angle) is given by the maximum wave-vector that can propagate in the superstrate:

$$k_{inc} = k_0 n_{sub} \sin \theta_{crit} = k_0 n_{sup} \qquad (1.4)$$

The critical angle θ_{crit} is straightforwardly given by

$$\theta_{crit} = \arcsin \frac{n_{sup}}{n_{sub}} = 41.81° \qquad (1.5)$$

Beyond the critical angle light undergoes Total Internal Reflection (TIR) regardless of the polarization state.

Due to the conservation of the electric field component parallel to the interface, the electromagnetic field in the superstrate does not fall abruptly to zero when $k_{inc} \geq k_0 n_{sup}$, but the solution in the superstrate is such that E_z is imaginary thus leading to an evanescent electromagnetic field exponentially damped along z. In the region where the forbidden band of the 1DPC (marked by a dashed black line) overlaps the TIR region light undergoes double reflection.

We can figure out to exploit the double reflection mechanism both towards the superstrate because of TIR and towards the ML because of Bragg reflection to guide light at the truncation interface of the 1DPC. To this aim we need to engineer the ML so that a defect state appears at the interface between the 1DPC and the superstrate. In the ideal case of a semi-infinite 1DPC, the reflectivity within the forbidden band is 1. In real systems with a finite number of layers, an evanescent tail crossing the DBR allows to couple the far-field radiation to BSW whenever the substrate has a refractive index higher than the effective index of BSW [20] in the well-known Kretschmann-Rather configuration or by employing immersion optics.

The invariance along the in-plane (x, y) directions of the photonic structure combined with the double confinement along the out-of-plane (z) direction ensures the condition for guiding light along the surface of the 1DPC. The electromagnetic surface modes obtained are known as Bloch Surface Waves (BSWs). In order to discuss some features of the BSWs, we can consider a specific design of ML, sketched in Fig. 1.5. The 1DPC is a stack of Ta_2O_5 ($n_h = 2.08 + i2e - 4$ at $\lambda = 600$ nm) and SiO_2 ($n_l = 1.45 + i2e - 4$ at $\lambda = 600$ nm) thin films deposited on a glass coverslip (BK7, $n_{sub} = 1.5$ at $\lambda = 600$ nm, thickness 150μm) by plasma ion-assisted deposition under high vacuum conditions (APS904 coating system, Leybold Optics). The high and low refractive index layers have a thickness of 70 nm and 210 nm respectively and there are 3 pairs of layers, whereas the fourth pair is terminated with a layer of SiO_2 180 nm thick. The complex refractive indices have been determined by ellipsometric measurements, and a detailed reference about the estimation of the values can be found in *Michelotti et al., Opt. Lett. 38, 616-18 (2013)* [21].

The reflectance map associated with the described photonic structure has been calculated by the well-known Transfer Matrix Method (TMM). The TE reflectance map (Fig. 1.3b) calculated as function of $(\lambda, k_T/k_0)$ shows that beyond the light line and in the visible range, a narrow dispersed dip in reflectance appears. The dip is

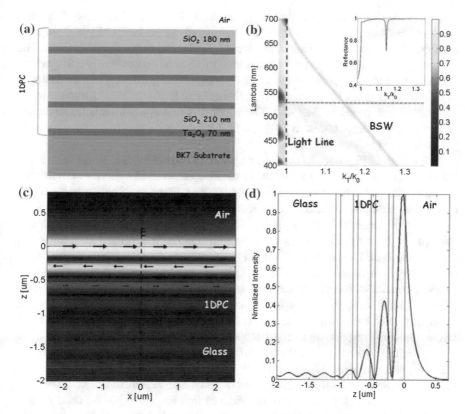

Fig. 1.5 **a** Cross-sectional view of the 1DPC grown on a glass substrate. **b** Calculated TE reflectance map as function of λ and θ. The inset shows an intensity profile extracted along the *green dashed line* (at $\lambda = 532$ nm). **c** Electric field distribution across the multilayer of the BSW mode calculated at $\lambda = 532$ nm. The *black arrows* indicate the in-plane orientation of the electric field. **d** Normalized intensity profile of the electric field extracted along the *black dashed line* in (**c**)

associated with the transfer of energy from the incoming plane wave to the surface mode, that propagates along the truncation interface with a wavevector that is given by the following relation:

$$k_{BSW} = k_0 n_{sub} sin(\theta_{BSW}) \tag{1.6}$$

where k_0 is the free space wavevector and θ_{BSW} is the BSW coupling angle.

As clearly observable in Fig. 1.5b, c, and d respectively, BSWs are characterized by a well defined dispersion relation and polarization and by a field distribution exhibiting the maximum of intensity at the 1DPC/dielectric interface. Because of the above mentioned reasons, BSWs can be considered as the photonic counterpart of Surface Plasmon Polaritons (SPPs), although the physics is completely different: in the case of a metallic surface, the localization of the field at the metal/dielectric interface is due to a negative dielectric constant in the metal [22], while BSWs localization occurs because of interference effects in the 1DPC [23].

Due to the phenomenological analogy among SPPs and BSWs, many of the concepts and methods that apply to SPPs have been implemented with BSWs. Compared to SPPs, BSWs exhibit some inherent advantages: the resonances' quality factors are indeed typically higher due to very low ohmic losses, and consequently BSW exhibit typically longer propagation distances [24]. Moreover, multilayers can be designed to sustain BSW in a wide spcetral range, from UV [25, 26] to IR [27], and the mode polarization can be selected as well. The possibility of employing functional dielectric materials in the ML design [28] can be further considered as an advantages in particular applications.

In the inset of Fig. 1.5b, a cut of the map at fixed wavelength ($\lambda = 532$ nm, along the green dashed line), shows that the resonance centered at $k_{BSW}/k_0 = 1.139$ has a Full Width Half Maximum (FWHM) of 0.006 in terms of effective index of the mode, corresponding to $0.53°$. The width of the resonance is ultimately associated to the overall losses and, consequently, to the decay length of the mode. To check the validity of our calculations, we compared the TMM results with a finite element method (FEM) model. In this model a cross section of the 1DPC is considered (Fig. 1.5c) and a modal analysis of the structure has been performed by imposing periodic conditions (continuity) at the left and right boundaries, and scattering condition with null incident electric field at the top and bottom boundaries. Among the different modes sustained by the structure, we found a TE polarized mode (black arrows show the orientation of the electric field) with the maximum of intensity in close proximity with the truncation interface (shown in Fig. 1.5). Such mode has an effective index (at $\lambda = 532$ nm) of 1.138, which is in very good agreement with the value found with the TMM method. A cross section of the electric field distribution along the structure (Fig. 1.5d) puts in evidence that the mode has its maximum in close proximity to the truncation interface and that it is exponentially decaying in the outer dielectric medium. Within the multilayer, the envelop of the fringes has an exponentially decaying profile as well, while the radiation reaching the substrate (leakage radiation) shows a propagative behavior without attenuation.

Due to the evanescent nature of BSWs, it is not possible to directly couple freely propagating light in air to BSW both in excitation and in collection. Since the BSW momentum is larger than the free-space momentum, a strategy to increase the momentum of the incident light is required. A simple way, widely used also in plasmonics, is to couple the 1DPC with a prism, as in the Rather Kretschmann or in the Otto configurations. By using a near field probe it is also possible to excite the evanescent modes [29] as well as to collect the evanescent optical fields directly on the flat surface of 1DPC, providing the amplitude [30] and eventually the phase [31] distributions of light.

Here, we make use of a customized Leakage Radiation Microscope (LRM). The setup is sketched in Fig. 1.6: the microscope is equipped with two sources, namely an halogen lamp and a collimated Nd:Yag doubled frequency laser ($\lambda = 532$ nm). From the bottom side, the laser beam is expanded and collimated by a beam expander and then spectrally cleaned at 532 nm by a Laser Line filter (MaxLine filter 532. Semrock). The collimated beam passes through a beam splitter and is then focused on the sample by an oil-immersion objective (Nikon APO TIRF 100x) with high

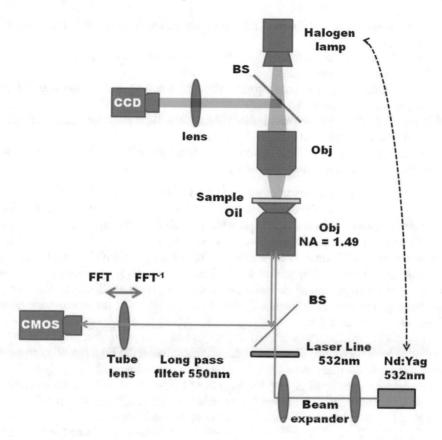

Fig. 1.6 Schematic view of the customized Leakage Radiation Microscope employed in the experiments

Numerical Aperture ($NA = 1.49$). On the other side the microscope is equipped with another objective and a white halogen lamp. The white lamp can be also used on the bottom side. Both sides are equipped with an imaging sensor (a CCD camera the top side, an RGB CMOS camera (Thorlabs DCC1645C) the bottom side. The working distance of the high NA objective is of the order of few hundreds of μm, thus imposing a constrain on the maximum thickness of the substrate (in our case we used 150 μm thick glass coverslips). The microscope is equipped in such a way that both illumination and collection can be performed either in transmission or in reflection mode. So the possible configurations are:

- Laser illumination from bottom side, laser collection from bottom side (laser reflection mode)
- Laser illumination from bottom side, fluorescence collection from bottom side (epi-fluorescence mode)

- Laser illumination from bottom side, laser collection from top side (laser transmission mode)
- Laser illumination from bottom side, fluorescence collection from top side (fluorescence transmission mode)
- White light illumination from top side, white light collection from top side (white light reflection mode (top))
- White light illumination from bottom side, white light collection from top side (white light transmission mode)
- White light illumination from bottom side, white light collection from bottom side (white light reflection mode (bottom))

From the bottom side, the microscope is capable of imaging both the direct plane (DP) and the back focal plane (BFP) of the collection objective, i.e. the spatial Fourier Transform of the DP. In this way, it is possible to obtain both the spatial and the angular distribution of the optical fields collected with the oil immersion objective [32]. The radiation collected is imaged via a tube lens onto the RGB CMOS camera. The air side and the sample can be moved freely from the objectives by means of a piezo drive along the three Cartesian directions ($100 \, \mu m \times 100 \, \mu m \times 20 \, \mu m$). Depending on the experimental requirements, different configurations of the microscope have been used in the following.

In order to experimentally test our multilayer and observe the coupling of far-field radiation with BSWs, we illuminated the sample through the high NA objective oil contacted with the glass substrate with a quasi-collimated white halogen lamp. The white light is spectrally filtered by the laser line filter transmitting light in the spectral range $\lambda = 532 \pm 1$ nm. Incoherent light improves the image quality, avoiding the problems of speckles that would affect a coherent light image.

The image in Fig. 1.7a indicates the reflected light intensity when the unpolarized incoherent light is employed. The reflection angle θ is such that $n_{obj} sin\theta = \sqrt{n_{obj}^2 sin^2(\theta_x) + n_{obj}^2 sin^2(\theta_y)}$ where x and y indicate the two *in-plane* directions, while n_{obj} is the refractive index of the collecting objective (oil-contacted with the prism by a matching index oil). The leakage output directions are related to the corresponding wavevector components as $k_{x,y} = k_0 n_{obj} sin(\theta_{x,y})$, where $k_0 = 2\pi/\lambda$ is the wavevector modulus in vacuum. The external circle has a diameter corresponding to the NA of the collecting optics. The BFP image shows a significant increase of the reflected light intensity when the TIR condition is achieved, i.e. when $\sqrt{(k_x/k_0)^2 + (k_y/k_0)^2} = n_{obj}\sqrt{sin^2\theta_x + sin^2\theta_y} = 1$ (light cone in air). For leaking directions comprised within the light line in air, the 1DPC is basically transmissive and only a small fraction of light is reflected according to the photonic band structure of the multilayer. In the TIR regime, the reflected light is imaged in an annular region comprised between an inner radius defined by the light cone in air and an external radius defined by the objective NA. In such an annular region, the light is distributed homogeneously, since it is completely reflected. The only exception is at a radius $\sqrt{(k_x/k_0)^2 + (k_y/k_0)^2} \approx 1.15$, where a narrow low-reflectivity dip is found [33]. A more detailed analysis can be performed by looking at the cross-sectional distribution of the BFP intensity along the horizontal dashed line in Fig. 1.7, and by

Fig. 1.7 BFP image (on the *top*) of the NA = 1.49 objective illuminating a flat 1DPC with incoherent, unpolarized light (filtered at λ = 532 ± 1 nm). On the *bottom*, the cross-sectional intensity profile extracted along the *red dashed line* (*solid green line*) and calculated reflectivity profile for an s-polarized plane wave at λ = 532 nm (*solid blue line*). Reprinted from [33]

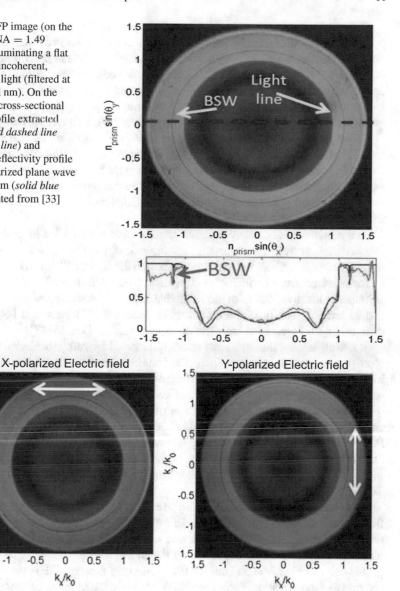

Fig. 1.8 Back Focal Plane images of incoherent light filtered at λ = 532 nm and polarized along x (image on the *left*) and y (image on the *right*). The images show dips of reflectance associated to the coupling of light with the surface mode

comparing it with the angular reflectivity calculated at λ = 532 nm with the TMM method. The low depth of the dip can be explained by recalling that the incident light is unpolarized, while in this case BSWs are allowed in TE-polarization only.

Figure 1.8 confirms that the BSW mode is polarized. When a polarization filter is added on the illumination path, the dark circle turns into two half moons, and only radiation azimuthally polarized couples to BSW, while radially polarized light is reflected. More details about the above mentioned characterization of the 1DPC can be found in [33]. The natural filtering operated by the photonic structure in the k−space provides also an interesting opportunity to generate evanescent complex fields such as Bessel beams or optical vortices on the multilayer surface [34].

BSWs can be considered as an effective alternative to Surface Plasmon Polaritons (SPPs) as information carriers in applications that can take advantage from an enhanced density of photonic states at the interface between the substrate and the external environment, such as sensing on planar supports or surface photonic circuitry. The inherent advantages of BSWs over SPPs make them a suitable candidate in such applications.

In the last decades, many research groups have put in a lot of efforts into the manipulation of SPP, by employing refractive or diffractive surface structures [35–42]. Similarly to SPPs on flat and structured metallic films, BSWs can be manipulated by means of refractive or diffractive structures. By exploiting the BSW effective index shift due to local variations of the refractive index at the surface of the 1DPC, BSW can be laterally confined and guided [24], reflected [43], refracted [44], in-plane diffracted and out-coupled from the multilayer surface [33, 45–47].

It is well known that SPPs are strongly affected by refractive index changes in close proximity to the surface. Such effect has been extensively studied and it is now implemented also in commercial sensing devices. The basic principle of SPR based sensors is that a variation of the refractive index in the external dielectric medium (e.g. air or water) results in a variation of the effective index of the mode, i.e. a shift of the SPP dispersion relation [48]. The same effect has been observed when BSWs are considered. Many sensing schemes based on the dielectric loading effect have been proposed in the last years, both to monitor the refractive index changes in the environment [49] and to monitor binding events occurring at the surface of a 1DPC [50].

Such effect can be exploited to fabricate 2D refractive elements that would allow for in-plane manipulation of BSWs. In Fig. 1.9 the electric field distributions corresponding to the two BSW modes, one on the bare surface and the other within a surface polymeric ridge deposited on the multilayer, are shown. The reflectance profile calculated at $\lambda = 532\,nm$ for the Transverse Electric (TE) polarization, also referred as s-polarization, shows the BSW effective index in the case of a bare multilayer (black line) and in the case of the same multilayer coated with a 30-nm-thick polystyrene (PPST, n = 1.55) layer (red line). BSW on bare multilayer has a calculated effective index equal to 1.14, while the BSW peak position on the PPST coated multilayer appears at an effective index equal to 1.19, thus confirming the red-shift of the mode in presence of a thin dielectric coating (Fig. 1.10). The red-shift can be exploited to confine the bsw mode or to affect its propagation by modifying its wavefront.

BSW mode on coated surface   BSW mode on bare surface

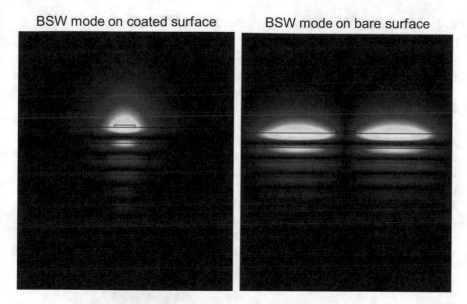

Fig. 1.9 Cross-sectional view of the electric field distribution of the BSW mode within the polymeric structure (on the *left*) and on the bare multilayer (on the *right*)

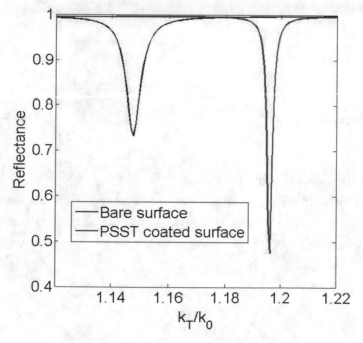

Fig. 1.10 Reflectance profile calculated by TMM method at fixed wavelength ($\lambda = 532$ nm) on the bare surface (*black line*) and on the coated surface (*red line*)

1.2   Ultrathin Refractive Structures for 2D Optics

Surface relief polymeric structures like thin biconvex structures or ridges have been fabricated on the multilayer by direct laser lithography on a positive photoresist followed by plasma assisted PPST deposition (30 nm) and subsequent lift-off in acetone. The experimental setup employed to characterize the refractive properties of the PPST structures is depicted in Fig. 1.11. The glass substrate hosting the multilayer is oil contacted to a prism, according to the Kretschmann configuration. A doubled frequency Nd:YAG CW laser beam, whose lateral size and divergence can be adjusted by means of a beam expander and a diaphragm, impinges on the 1DPC through the prism. The incident angle θ_{BSW} is set in such a way that the transverse wavevector k_T of BSW at $\lambda = 532$ nm can be matched through the relationship $k_T = n_{prism} \sin(\theta_{BSW})$, where n_{prism} is the glass-prism refractive index.

The intensity distribution of BSW on the surface of the 1DPC can be observed in several ways. A direct way consists in collecting the evanescent fields by means of

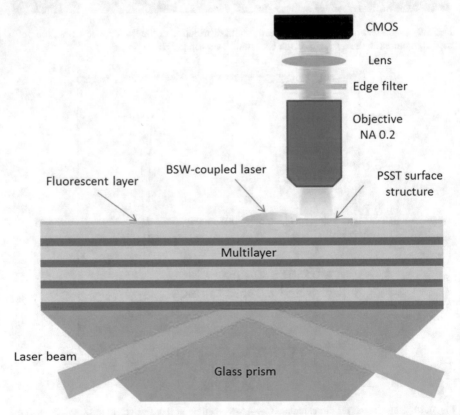

Fig. 1.11 Sketch of the experimental setup employed to observe the fluorescent trace of BSW coupled through a prism oil contacted in the Kretschmann configuration. Illumination is provided by a quasi-collimated laser beam

a near-field probe [27]. Although this method can provide both amplitude and phase distribution of the field on the surface, it is time consuming and requires a dedicated apparatus. Here a simpler strategy is implemented: a fluorescent dye is used as probe for revealing the BSW coupled radiation intensity, in such a way that the fluorescence trace of the underlying excitation field can be obtained, thus mapping the BSW spatial distribution.

In order to obtain an homogeneous fluorescent layer, we incubated protein A conjugated with Alexa Fluor 546 (PtA-Alexa 456) on the multilayer surface for 20 min. and then rinsed with phosphate buffered saline (PBS) buffer. In this way, fluorophores act as a probe for the BSW near-field intensity, thus allowing a simple wide field fluorescence imaging from the air side with a CMOS camera. The main drawback of this approach is represented by the absorption introduced by the fluorophores on the surface that affects the decay length of BSW, as explained in detail in the following section.

In order to evaluate the focusing effect and to have a realistic comparison between experimental data and modeling, an evaluation of the BSW decay length is required. Starting from the TMM calculations, a theoretical estimation of the decay length can be obtained by extracting the Full Width Half Maximum (FWHM) of the dip in reflectance, according to the following relation [51]:

$$L_D = \lambda_{BSW}/(\pi\sigma) \tag{1.7}$$

where λ_{BSW} is the effective BSW wavelength, while σ is the FWHM of the resonance. In this case, the width of the resonance is determined by two lossy channels: the intrinsic losses of the materials employed and the leakage radiation.

The first one depends on the quality of the dielectric layers and on the defects of the surface producing scattered light. An ellipsometric analysis of the dielectric thin films revealed no detectable absorption. Since in TMM calculations an absorption coefficient is required to observe the dip, we set the imaginary part of the refractive indices at a value of 2×10^{-4}, which is basically the lower limit of detection of the employed instrument. Moreover, in previous papers where the role of absorption has been discussed, such value has been successfully used to reproduce experimental results in analogous multilayer structures [21].

The effect of leakage radiation on the mode losses is mostly related to the 1DPC design. The amount of radiation that tunnels through the multilayer depends on the reflectivity of the bragg structure for the specific energy-momentum pair given by the BSW dispersion relation. When the elementary cell (i.e. the high-low refractive index layers pair) of the multilayer is fixed, the reflectivity only depends on the numbers of pairs. Increasing the number of pairs frustrates the tunneling of radiation from the BSW to the substrate, reducing the overall losses of the mode and therefore narrowing the resonance (Fig. 1.12).

The slight red-shift observed when the number of layers is increased can be explained by considering the field distribution of the mode: when the number of layers increases, the percentage of the energy confined inside the multilayer increases with respect to the percentage of energy distributed in the external evanescent tale,

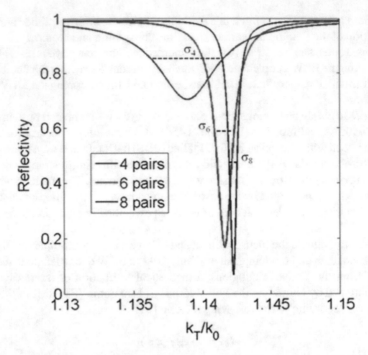

Fig. 1.12 Reflectance profile as function of the normalized incident wavevector calculated at $\lambda = 532$ nm for different multilayers with the same elementary cell but different number of layers

thus increasing the effective refractive index of the mode. It has to be noticed that a narrower resonance, corresponding to a higher Q-factor, means that the photonic mode is less coupled with the external environment, and it is therefore more diffi-cult to couple far field radiation to the mode itself for example by employing the Kretschmann configuration or oil immersion optics.

By applying Eq. 1.7, the theoretical decay length on the bare multilayer employed in the experiment (4 pairs) is 34 μm, while it is about one order of magnitude larger for the 8 pairs ML. In order to obtain an experimental estimation of the decay length, we launched a BSW on the flat surface (inset in Fig. 1.13) and measured the fluorescence intensity. In Fig. 1.13, the experimental data (blue line) were fitted with the following relation:

$$I(y) = I_0 e^{(-y/L_D)} \tag{1.8}$$

where I_0 is the measured initial fluorescence intensity and L_D is the fitting parameter corresponding to the BSW decay length. The best fit (red dashed line) has been obtained for $L_D \approx 28$ μm. The difference between the experimental data and the theoretical value has been attributed to the absorption caused by the fluorescent

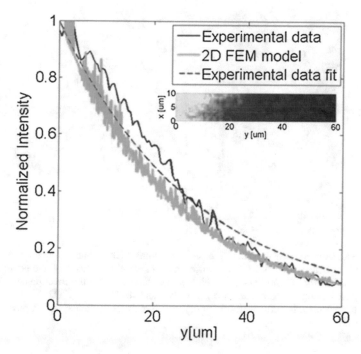

Fig. 1.13 Measured (*blue*) and calculated (*green*) normalized intensity profile of fluorescence excited by the BSW coupled at $y = 0$. The *red dashed line* has been obtained by fitting the experimental data with Eq. 1.8. In the inset, the fluorescent trace of the BSW as imaged on the CMOS camera reported in false colors

layer. In addition, clots of protein act as scatterers on the surface, thus increasing the overall losses [52].

In order to obtain a realistic comparison with a 2D FEM model, such decay length has to be reproduced in an effective medium model [27]. The 2D medium mimicking the surface of the 1DPC is therefore modeled with a refractive index in which the real part corresponds to the effective index of the BSW mode, as derived from the TMM calculations, while the imaginary part has been set according to the following relation:

$$k = \lambda_{eff}/(4\pi L_D) = 1.8 \times 10^{-3}. \tag{1.9}$$

It has to be pointed out that such absorption coefficient is not related to the absorption of the materials, but it is just a fictitious parameter allowing a direct comparison between the propagation of BSW on the surface of a structured medium such as the 1DPC and an effective homogeneous medium. The green line in Fig. 1.13 corresponds to the intensity of a plane wave ($\lambda = \lambda_{eff}$) propagating into such an

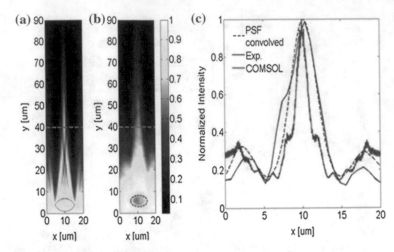

Fig. 1.14 **a** Calculated BSW focusing by means of a 30-nm-thick lens having a radius of curvature of 6 μm and diameter 10.4 μm. **b** Fluorescence image of BSW as focused by the polymeric lens (marked with the *black dashed line*). The *blue dashed lines* indicate the focal plane. **c** Intensity profile extracted along the *blue lines* in (**a**) (*red line*) and (**b**) (*blue line*). The *black dashed line* represents the calculated profile convolved with the PSF of the collection system. Copyright OSA. Reprinted from [52]

homogeneous layer. The good agreement between the model and the experiment confirms the validity of the above described approach.

The model implemented is a 2D Finite Element Method (FEM) model in which a plane wave propagates in an homogeneous medium characterized by a refractive index real part equal to the effective index of the BSW mode as calculated from the TMM method, and an absorption coefficient derived from the measured decay length.

The polymeric planar lens is a biconvex lens with a radius of curvature of 6 μm and a diameter of 10.4 μm. The expected focal plane is located at about 35 μm from the lens. The 2D effective index model shows the intensity distribution reported in Fig. 1.14a, where the focusing effect of the bi-dimensional lens is visible. Figure 1.14b shows the experimental observation of the focusing effect, confirming that the BSW-coupled radiation converges toward a point located at about 35 μm from the center of the lens. Concerning the intensity, the field enhancement expected by the BSW focusing cannot be fully evaluated due to the high losses affecting the BSW propagation.

In order to compare the model prediction with the experimental results, we extracted the cross-sectional intensity distribution in the focal region, i.e. at about 35 μm far from the center of the lens. The peak associated to the calculated intensity distribution has a Full Width Half Maximum (FWHM) of about 2 μm. The experimental cross section shows a much broader peak, but here the spatial resolution of the collection optics needs to be taken into account. To this aim, we computed the convolution of the calculated profile with a Gaussian function (roughly mimicking the point spread function (PSF) of the collection objective) whose variance is given by the Rayleigh resolution limit:

$$\sigma^2 = \lambda/(2NA) \approx 1.43\,\mu m. \tag{1.10}$$

where $\lambda = 570\,nm$ is the peak fluorescence emission wavelength and NA $= 0.2$ is the numerical aperture of the collection optics. After convolution, the FWHM of the calculated field intensity profile in the focal plane is about $4.3\,\mu m$, which is in rather good agreement with the measured intensity profile ($FWHM = 5 \pm 1\,\mu m$). A limiting factor in the estimation accuracy is given by the image pixelization of the CMOS camera employed to record the fluorescent image. Beside this, it can be noted that the fluorescence intensity is weaker on the polymeric lens. This is attributed to the anti-fouling behavior of the PPST [53].

Two-dimensional refractive elements can also serve as waveguides for the BSW-coupled radiation. In order to inject light into such surface waveguides a lens is placed in front of a thin planar ridge. The terminal end of the ridge is about $35\,\mu m$ far from the lens, i.e. approximately within its focal plane. The waveguide is $100\,\mu m$ long, $5\,\mu m$ wide, and $30\,nm$ thick and is made of PPST, as is the lens. The FEM analysis shows that the BSW, focused by the lens, is actually injected into the ridge and guided within it (Fig. 1.15a). The guiding behavior of the polymeric ridge is confirmed by the experimental observation, showing that the effective index variation produced by a 30 nm PPST layer is large enough to guide the BSW in a straight direction. (Fig. 1.15b)

As previously highlighted, the anti-fouling property of the PPST structure inhibits the adhesion of PtA Alexa on the waveguide structure, thus reducing the fluorescence intensity thereon observed. When we consider a longitudinal cross-sectional profile of intensity along the guide (Fig. 1.15c), we observe a sharp change in the fluorescence intensity (i.e. of the PtA-Alexa concentration) at both terminal ends of the ridge, as indicated. Within the waveguide, the intensity profile (integrated over few pixel rows along the x axis) exhibits an exponentially decaying profile. The decay constant is extracted by fitting the experimental intensity profile with Eq. 1.8. Best fit, in this case, is obtained for $L_D = 83.3\,\mu m$, that is roughly four times larger than the propagation length observed on the bare surface.

This value is in good agreement with the theoretical value of $83.7\,\mu m$ obtained by the TMM method for a multilayer coated by a uniform 30-nm-thick PPST layer. The resonance for the considered multilayer in presence of a thin dielectric coating is indeed much narrower with respect to the resonance on the bare substrate (see Fig. 1.10). The PPST anti-fouling property limits the role of absorption due to the fluorescent probes. In the effective medium model, we reduced the absorption coefficient in the waveguide by a factor of four (i.e. we set an absorption coefficient of 4.38×10^{-4}). By this way we found a good matching between calculations and experiment (see Fig. 1.15a, b).

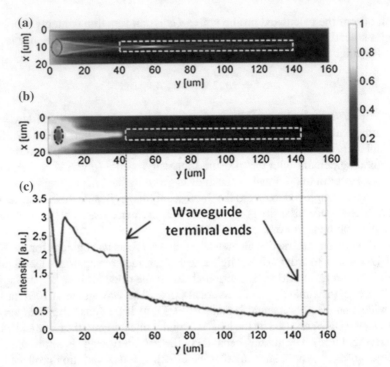

Fig. 1.15 a Calculated and **b** measured intensity distribution of a BSW injected into a polymeric ridge waveguide (contour, *dashed white line*) by means of a 2D lens. **c** Experimental intensity profile along the y axis extracted from **b**. The vertical *dotted lines* indicate the waveguide terminal ends. Inside the waveguide the profile is well fitted by an exponentially decaying function (*dashed red line*). Copyright OSA. Reprinted from [52]

1.3 Resonant Coupling Assisted by Surface Diffractive Structures

Because of their evanescent nature, no direct coupling between BSWs and propagating photons in the outer medium (air in the present case) can occur. However, a surface grating might provide an additional momentum large enough to couple the incident light to BSWs through, for example, first-order diffraction [33]. In the past, it has been shown that a sub-wavelength grating can back-reflect BSWs, thus producing an energy band-gap in the dispersion relation of the BSW [43]. In the present case, a linear grating with spatial period $\Lambda_g = 520\,$nm is patterned onto the 1DPC (Fig. 1.16a). The grating has been fabricated by FIB lithography on the surface of the 1DPC.

In order to couple incident radiation to BSWs, the grating period should be tailored according to the in-plane Bragg's law:

$$k_{diff} = k_{inc} \pm G \tag{1.11}$$

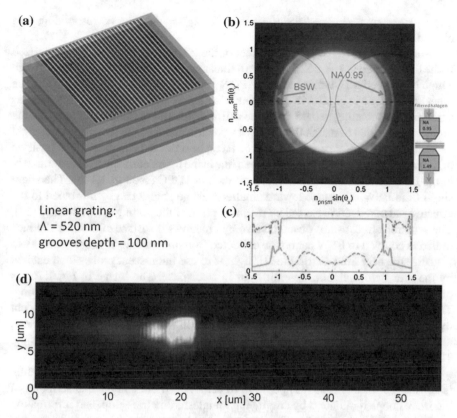

Fig. 1.16 **a** Sketch of the one-dimensional photonic crystal with a linear grating patterned on the surface. The *top* layer of the 1DPC is a SEM image of the grating. **b** LRM image of the BFP when the grating is enlighted from the air side with a NA — 0.95 objective. The light comes from a halogen lamp and is filtered with a Laser Line Filter ($\lambda = 532$ nm). **c** Intensity profile extracted from **b** along the *red dashed line* (*green solid line*). For comparison, also the intensity profile extracted form Fig. 1.7 (*green dashed line*) is reported. **d** Wide field fluorescence image of the BSW launched by the grating on the flat surface

where k_{diff} is the momentum of the diffracted beam, $k_{inc} = k_0 sin(\theta_{inc})$ is the wavevector component of the incident beam parallel to the surface and $G = 2\pi/\Lambda_g$ is the grating momentum. When k_{diff} equals k_{BSW}, coupling between far-field radiation and BSW can occur [33]. For the multilayer employed here the dispersion relation shows that the coupling angle for $\lambda = 532$ nm is 49.5 deg. By applying Eq. 1.6, the effective wavelength of BSW is found to be $\lambda_{BSW} = 2\pi/k_{BSW} = 466$ nm. The mismatch between the effective wavelength of radiation coupled to BSW and the grating period can be compensated by using an objective with a NA high enough to ensure that the incident light has a wide spread of k-vectors including the ones that satisfy Eq. 1.11. In the present case, the incident angle required to couple radiation impinging on the grating from air is $\theta_{inc} = arcsin((k_{BSW} - G)/k_0) = 6.7$ deg, corresponding to a $NA \simeq 0.11$.

In order to observe the coupling effect of the grating, we focused an unpolarized white light filtered at $\lambda = 532 \pm 1$ nm on a linear grating. The light is focused through a high NA objective (Olympus MPlanAPON 100x, NA = 0.95) operating in air. The inner bright circle in Fig. 1.16b corresponds to directly transmitted light, and thus has a radius corresponding to 0.95. The first diffraction order of the grating provides the momentum needed to match the BSW wavevector, and thus the two bright arcs right beyond the light line appear. In Fig. 1.16c it can be observed that the angular position of the two bright arcs along the red dashed line in the figure correspond approximately to the BSW reflectivity dips observed on the flat 1DPC (see Fig. 1.14b). The slight shift of the BSW resonance toward smaller leakage angles can be attributed to the effective index reduction caused by the inscription of the grating on the surface [33]. The coupling effect can be observed also by imaging the surface of the 1DPC. Since radiation coupled to BSW cannot be collected from the air side, a simple strategy to visualize the near field distribution of light is to use fluorescent probes that, excited by the near field radiation, re-emit light in free space. The image in Fig. 1.16d is obtained by focusing a laser spot ($\lambda = 532$ nm) on a portion of a linear grating. The surface of the 1DPC is covered with an homogeneous layer of a fluorescent dye (Rhodamine B). The fluorescent trace of the BSW, imaged on a CMOS camera, allows to easily visualize the near field distribution of light on the surface of the 1DPC.

The phenomenon described above can be also observed when coherent radiation is considered. In this case it is possible to achieve a more complete description of the BSW-coupled radiation by employing an interferometric setup, and information about the phase and amplitude distribution of light coupled to the surface modes can be retrieved in addition to the intensity pattern [54].

An implementation of the leakage radiation interference microscope (LRIM) is schematically shown in Fig. 1.17, and a detailed description of its working principle can be found in *M. S. Kim et al., Opt. Lett. 37 (3), 305–307, 2012.* Briefly, a collimated beam (CW frequency doubled Nd:YAG laser at $\lambda = 532$ nm) is splitted by a Polarizing Beam Splitter (PBS). The polarization of the two beams can be further controlled by two Glenn-Taylor polarizers (G-T polarizers).

The first beam (object beam) is focused on the sample by an objective (NA = 0.5). The effective NA of the illumination beam can be adjusted by means of a diaphragm placed at the entrance of the objective pupil. Light emerging from the sample is collected by an oil-immersion objective (NA = 1.4) oil contacted with the sample substrate. The collected fields are imaged on a CCD camera via a tube lens.

On the other path, the reference beam is reflected by a piezo-actuated mirror that can be moved in steps of a quarter of the wavelength and is then sent on the CCD, thus interfering with the object beam. The five interference patterns obtained are post-processed with an algorithm based on the Schwider-Hariharan method [55, 56], through which it is possible to retrieve both amplitude and phase distributions of the collected optical fields. The experimental stage is such that the illumination objective can be moved along the z direction with respect to the sample, and the

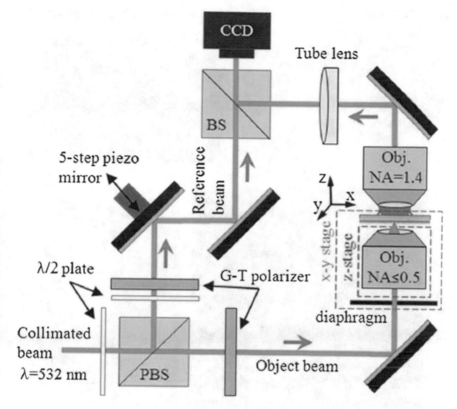

Fig. 1.17 Schematic view of the leakage radiation interference microscope. Copyright OSA. Reprinted from [54]

sample together with the illumination objective can be moved along x, y with respect to the collection objective, as schematically depicted in Fig. 1.18a.

By means of a Bertrand lens inserted in the collection optical path before the tube lens, a BFP image proportional to the Fourier transformation of the direct plane image [57] can be produced on the CCD camera. By looking at the BFP image it is indeed easier to separate the directly transmitted light (contained within a ring whose diameter is proportional to the illumination NA), and the leakage radiation [54]. The BFP image of the oil immersion objective is reported in Fig. 1.18c. The external (green) circle corresponds to the maximum NA of the collection objective (NA = 1.4), while the inner (blue) circle defines the light line in air (corresponding to NA = 1). The directly transmitted light appears as a circle with radius NA = 0.15, corresponding to the particular NA of the illumination beam. When the object beam polarization is aligned along the x axis and the grating vector is parallel to the y axis, a bright arc appears beyond the light line, at a location $k_y/k_0 = 1.19 \pm 0.2$. This is highlighted in Fig. 1.18d, where a cross section of the BFP intensity distribution along the red dotted line in Fig. 1.18c is plotted. The bright arc is polarized along the x

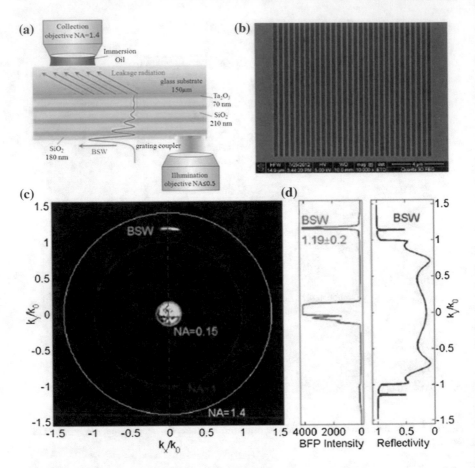

Fig. 1.18 a Detailed sketch of the photonic structure and the experimental arrangement used for sample illumination and leakage radiation collection. **b** scanning electron microscope image of the grating coupler for a Bloch surface wave (BSW). **c** BFP image of grating coupled BSW upon illumination with a NA = 0.15 beam. **d** Experimental (*green*) cross-sectional intensity profile of BFP along the *dashed white line* and calculated angularly resolved TE-polarized reflectance at λ = 532 nm revealing BSW resonances as indicated. Copyright OSA. Reprinted from [54]

direction, and it is associated to a BSW coupled by the grating and propagating on the multilayer surface. As a check, the calculated s-polarized angular reflectance profile at λ = 532 nm presented in Fig. 1.18d (reveals the presence of BSW resonance dips at $k_y/k_0 = 1.15$, as previously shown in Fig. 1.5 on the ideal multilayer layout from design.

By removing the Bertrand lens from the collection path, a direct image of the surface is produced on the CCD camera (Fig. 1.19a). The image of the BSW launched by the grating is obtained by collecting the leakage radiation through the oil immersion objective. In order to reduce the directly transmitted light, that in the direct plane

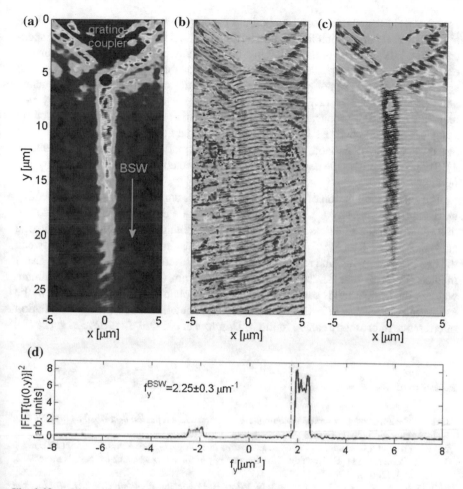

Fig. 1.19 **a** Direct plane image of BSW launched from a grating coupler. **b** Corresponding BSW phase distribution retrieved. **c** Real part of the obtained BSW complex field. **d** | *FFT* |² profile of the measured complex field along the propagation direction of BSW. Copyright OSA. Adapted from [54]

image cannot be separated from the leakage radiation, the collection objective is placed almost completely out of the field of view of the illumination objective. In these conditions, we performed a complete amplitude/phase analysis of the BSW-coupled radiation through the five-step interferometric technique. An exemplary result is presented in Fig. 1.19. Specifically, Fig. 1.19b shows the BSW amplitude distribution when the grating, placed at the top border of the image is directly illuminated. The BSW-coupled radiation propagates along the truncation interface according to the green arrow.

Figure 1.19c exhibits the BSW phase distribution, wherein the BSW propagation is revealed by a periodic sawtooth phase profile, as observed elsewhere [24]. The full complex field distribution can be obtained by combining the amplitude and phase maps. The real part of such field is illustrated in Fig. 1.19c. The spatial Fourier spectrum (Fig. 1.19d) can be obtained if a fast Fourier transformation (FFT) is performed on the complex field recovered along the propagation direction. The obtained spatial spectrum shows two peaks symmetrically placed with respect to the zero frequency and centered at $f_y^{BSW} = (2.25 \pm 0.3)\,\mu m^{-1}$. The peak with positive frequency corresponds to a BSW propagating in the positive y direction and is characterized by a higher amplitude because it is contained within the field of view of the collection objective. On the other hand, the peak exhibiting a smaller amplitude is associated to a BSW propagating in the negative y direction, collected essentially from the grating region.

The spatial modulation of the two detected peaks corresponds the BSW effective wavelength, as given by $\lambda_{eff}^{BSW} = \mid 1/(f_y^{BSW}) \mid = 444.4 \pm 59.3\,nm$. A comparison of this value with the effective BSW wavelength as obtained by from the BFP image analysis in Fig. 1.8 as $\lambda_{eff}^{BSW} = 442.2 \pm 7.5\,nm$, revealing a good matching between the two measurements, and to the theoretical value calculated by means of the TMM method (Fig. 1.3) that is $\lambda_{eff}^{BSW} = \lambda_0/n_{eff} = 465\,nm$. The limited spatial domain wherein the optical fields can be collected may explain the large uncertainty in the estimation of the BSW effective wavelength. More details about the leakage radiation interference microscopy can be found in *Descrovi et al., Opt. Lett. 38. 3374, (2013)*.

References

1. L. Rayleigh, *On the remarkable phenomenon of crystalline reflexion described by prof. Stokes*, Phil. Mag. S. 5, 26, (1888).
2. P. Vukusic and R. Sambles, *Photonic structures in biology*, Nature 424, 852–855, (2003).
3. A. R. Parker and H. E. Townley, *Biomimetics of photonic nanostructures.* Nat. Nanotech. 2, 347–353, (2007).
4. J. D. Joannopoulos, S. G. Jhonson, J. N. Winn, R. D. Meade, *Photonic Crystals. Molding the flow of light.*, 2nd ed., Princeton Press, Princeton, NJ, (2008).
5. T. Baba, *Slow light in photonic crystals*, Nature Photonics 2, 465–473, (2008).
6. T. F. Krauss, *Slow light in photonic crystal waveguides*, J. Phys. D: Appl. Phys. 40, 2666–2670, (2007).
7. O. Painter, R. K. Lee, A. Scherer, A. Yariv, J. D. O'Brien, P. D. Dapkus, I. Kim, *Two-Dimensional Photonic Band-Gap Defect Mode Laser*, Science 284 (5421), 1819–1921, (1999).
8. A. A. Erchak, D. J. Ripin, S. Fan, P. Rakich, J. D. Joannopoulos, E. P. Ippen, G. S. Petrich and L. A. Kolodziejski, *Enhanced coupling to vertical radiation using a two-dimensional photonic crystal in a semiconductor light-emitting diode*, Appl. Phys. Lett. 78, 563 (2001).
9. P. Russell, *Photonic Crystal Fibers*, Science, 299 (5605), 358–362, (2003).
10. L. Zeng, Y. Yi, C. Hong, J. Liu, N. Feng, X. Duan, L. C. Kimerling and B. A. Alamariu, *Efficiency enhancement in Si solar cells by textured photonic crystal back reflector*, Appl. Phys. Lett. 89, 111111, (2006).
11. E. Yablonovitch, T. J. Gmitter, R. D. Meade, A. M. Rappe, K. D. Brommer, and J. D. Joannopoulos, *Donor and acceptor modes in photonic band structure*, Phys. Rev. Lett. 67, 3380, (1991).
12. P. Lalanne C. Sauvan and J. P. Hugonin, *Photon confinement in photonic crystal nanocavities*, Laser and Photonics Reviews 2 (6), 514–526, (2008).

13. J. D. Joannopoulos, P. R. Villeneuve, S. Fan, *Photonic crystals: putting a new twist on light*, Nature 386, 143–149, (1997).
14. S. Noda, M. Fujita, T. Asano, *Spontaneous-emission control by photonic crystals and nanocavities*, Nature Photonics 1, 449–458, (2007).
15. D. Englund, D. Fattal, E. Waks, G. Solomon, B. Zhang, T. Nakaoka, Y. Arakawa, Y. Yamamoto and J. Vuckovic, *Controlling the Spontaneous Emission Rate of Single Quantum Dots in a Two-Dimensional Photonic Crystal*, PRL 95, 013904, (2005).
16. P. Lohdal, A. F. van Driel, I. S. Nikolaev, A. Irman, K. Overgaag, D. Vanmaekelbergh and W. L. Vos, *Controlling the dynamics of spontaneous emission from quantum dots by photonic crystals.*, Nature 430, 654–657, (2004).
17. E. M. Purcell, Phys. Rev. 69, 681, (1946).
18. L. Novotny and N. F. Van Hulst, *Antennas for light*, Nat. Photon. 5, 83–90, (2011).
19. M. D. Leistikow, A. P. Mosk, E. Yeganegi, S. R. Huisman, A. Lagendijk, and W. L. Vos, *Inhibited Spontaneous Emission of Quantum Dots Observed in a 3D Photonic Band Gap*, Phys. Rev. Lett. 107, 193903, (2011).
20. F. Michelotti, B. Sciacca, L. Dominici, M. Quaglio, E. Descrovi, F. Giorgis and F. Geobaldo, *Fast optical vapour sensing by Bloch surface waves on porous silicon membranes*, Phys. Chem. Chem. Phys 12, 502–506, (2010).
21. F. Michelotti, A. Sinibaldi, P. Munzert, N. Danz, and E. Descrovi *Probing losses of dielectric multilayers by means of Bloch surface waves*, Opt. Lett. 38, 616–618, (2013).
22. H. Raether, *Surface Plasmons*, Springer-Verlag, Berlin (1988).
23. R. D. Maede, K. D. Brommer, A. M. Rappe and J. D. Joannopoulos, *Electromagnetic Bloch Waves at the surface of a photonic crystal* Phys. Rev. B, 44, 109601, (1995).
24. E. Descrovi, T. Sfez, M. Quaglio, D. Brunazzo, L. Dominici, F. Michelotti, H. P. Herzig, O. J. F. Martin and F. Giorgis, *Guided Bloch surface waves on ultrathin polymeric ridges*, Nano Lett. 10, 2087–2091 (2012).
25. N. Ganesh, I. D. Block and B. T. Cunningham, *Near ultraviolet-wavelength photonic-crystal biosensor with enhanced surface-to-bulk sensitivity ratio*, Appl. Phys. Lett. 89, 023901, (2006).
26. R. Badugu, J. Mao, S. Blair, D. Zhang, E. Descrovi, A. Angelini, Y. Huo and J. R Lakowicz, *Bloch Surface Wave-Coupled Emission at Ultra-Violet Wavelengths, J. Phys. Chem. C, 120 (50), 28727–34, (2016).*
27. L. Yu, E. Barakat, T. Sfez, L. Hvozdara, J. Di Francesco and H. P. Herzig, *Manipulating Bloch surface waves in 2D: a platform concept-based flat lens*, Light: Sci. Appl. 3, e124, (2013).
28. M. Ballarini, F. Frascella, F. Michelotti, G. Digregorio, P. Rivolo, V. Paeder, V. Musi, F. Giorgis and E. Descrovi *Bloch surface waves-controlled emission of organic dyes grafted on a one-dimensional photonic crystal*, Appl. Phys. Lett 99, 043302 (2011).
29. B. Hecht, H. Bielefeldt, L. Novotny, Y. Inouye and D. W. Pohl, *Local excitation, scattering, and interference of surface plasmons*, Phys. Rev. Lett. 77 (9), 1889, 1996.
30. T. Sfez, E. Descrovi, L. Yu, D. Brunazzo, M. Quaglio, L. Dominici, W. Nakagawa, F. Michelotti, F. Giorgis, O. J. F. Martin and H. P. Herzig, *Bloch surface waves in ultrathin waveguides: near-field investigation of mode polarization and propagation*, JOSA B, 27 (8), 1617–1625, (2010).
31. X. Wu, E. Barakat, L. Yu, L. Sun, J. Wang, Q. Tan, H. P. Herzig *Phase-sensitive near field Investigation of Bloch surface wave propagation in curved waveguides*. JEOS - RP, Europe, v. 9, oct. 2014. ISSN 1990-2573.
32. A. Drezet, A. Hohenau, D. Koller, A. Stepanov, H. Ditlbacher, B. Steinberger, F. R. Aussenegg, A. Leitner, J. R. Krenn, *Leakage radiation microscopy of surface plasmon polaritons*, Mat. Sci.Eng. B 149, 220–229, (2008).
33. A. Angelini, E. Enrico, N. De Leo, P. Munzert, L. Boarino, F. Michelotti, F. Giorgis and E. Descrovi, *Fluorescence diffraction assisted by Bloch surface waves on a one-dimensional photonic crystal*, New J. Phys. 15, 073002, (2013).
34. A. Angelini, *Resonant evanescent complex field on dielectric multilayers*, Opt. Lett. 40(24), 5746–5749, (2015).
35. W. L. Barnes, A. Dereux and T. W. Ebbesen, *Surface Plasmon subwavelength optics*, Nature 424, 824–830, (2003).

36. J.-M. Yi, A. Cuche, E. Devaux, C. Genet, and T. W. Ebbesen, *Beaming Visible Light with a Plasmonic Aperture Antenna*, ACS Phot. 1, 365–370, (2014).
37. T. Zentgraf, Y. Liu, M.H. Mikkelsen, J. Valentine and X. Zhang, *Plasmonic Luneburg and Eaton lenses*, Nat. Nanotech. 6, 151–155, (2011).
38. C. Zhao and J. Zhang, *Flexible wavefront manipulation of surface plasmon polaritons without mechanical motion components*, Appl. Phys. Lett. 98, 211108, (2011).
39. C. Zhao, J. Wang, X. Wu and J. Zhang, *Focusing surface plasmons to multiple focal spots with a launching diffraction grating*, Appl. Phys. Lett. 94, 111105 (2009).
40. C. Zhao and J. Zhang, *Binary plasmonics: launching surface plasmon polaritons to a desired pattern*, Opt. Lett. 34, 2417, (2009).
41. C. Zhao, Y. Liu, Y. Zhao, N. Fang and T.J. Huang, *A reconfigurable plasmofluidic lens*, Nat. Comm. 4, 2305, (2013).
42. G. M. Lerman and U. Levy, *Pin Cushion Plasmonic Device for Polarization Beam Splitting, Focusing, and Beam Position Estimation*, Nano Lett. 13, 1100–1105, (2013).
43. E. Descrovi, F. Giorgis, L. Dominici and F. Michelotti, *Experimental observation of optical bandgaps for surface electromagnetic waves in a periodically corrugated one-dimensional silicon nitride photonic crystal*, Opt. Lett. 33 (3), 243–245, (2008).
44. T. Sfez, E. Descrovi, L. Yu, M. Quaglio, L. Dominici, W. Nakagawa and F. Michelotti, *Two-dimensional optics on silicon nitride multilayer: Refraction of Bloch surface waves*, Appl. Phys. Lett. 96, 151101, (2010).
45. A. Angelini, P. Munzert, E. Enrico, N. De Leo, L. Scaltrito, L. Boarino, F. Giorgis and E. Descrovi, *Surface-Wave-Assisted Beaming of Light Radiation from Localized Sources* ACS Phot. 1, 612–617 (2015).
46. A. Angelini, E. Barakat, P. Munzert, L. Boarino, N. De Leo, E. Enrico, F. Giorgis, H. P. Herzig, C. F. Pirri, E. Descrovi, *Focusing and Extraction of Light mediated by Bloch Surface Waves*, Sci. Rep. 4, 5428, (2014).
47. M. Liscidini, J. E. Sipe, *Analysis of Bloch-surface-wave assisted diffraction-based biosensors*, JOSAB 26 (2), 279–289, (2009).
48. T. Holmgaard and S. I. Bozhevolnyi, *Theoretical analysis of dielectric-loaded surface plasmon-polariton waveguides*, Phys. Rev. B 75, 245405, (2007).
49. E. Descrovi, F. Frascella, B. Sciacca, F. Geobaldo, L. Dominici, F. Michelotti, *Coupling of surface waves in highly defined one-dimensional porous silicon photonic crystals for gas sensing applications*, Appl. Phys. Lett. 91, 241109 (2007).
50. S. Santi, V. Musi, E. Descrovi, V. Paeder, J. Di Francesco, L. Hvozdara, P. van der Wal, H. A. Lashuel, A. Pastore, R. Neier, H. P. Herzig, *Real-time Amyloid Aggregation Monitoring with a Photonic Crystal-based Approach*, Chem. Phys. Chem. 14, 83–3476, (2013).
51. R. Ulrich, *Theory of the Prism-Film Coupler by Plane-Wave Analysis*, J. Opt. Soc. Am. 60, 1337 (1970).
52. A. Angelini, A. Lamberti, S. Ricciardi, F. Frascella, P. Munzert, N. De Leo and E. Descrovi, *In-plane 2D focusing of surface waves by ultrathin refractive structures*, Opt. Lett. 39 (22), 6391–6394, (2014).
53. M. Ballarini, F. Frascella, N. De Leo, S. Ricciardi, P. Rivolo, P. Mandracci, E. Enrico, F. Giorgis, F. Michelotti, and E. Descrovi, *A polymer-based functional pattern on one-dimensional photonic crystals for photon sorting of fluorescence radiation* Opt. Express 20, 6703 (2012).
54. E. Descrovi, E. Barakat, A. Angelini, P. Munzert, N. De Leo, L. Boarino, F. Giorgis and H. P. Herzig, *Leakage radiation interference microscopy*, Opt. Lett. 38 (17), 3374–3376, (2013).
55. J. Schwider, R. Burow, K. E. Elssner, J. Grzanna, R. Spolaczyk, and K. Merkel, *Digital wavefront measuring interferometry: some systematic error sources* Appl. Opt. 22, 3421 (1983).
56. P. Hariharan, B. F. Oreb, and T. Eiju, *Digital phase-shifting interferometry: a simple error-compensating phase calculation algorithm* Appl. Opt. 26, 2504 (1987).
57. A. Berrier, M. Swillo, N. Le Thomas, R. Houdre', and S. Anand, *Bloch mode excitation in two-dimensional photonic crystals imaged by Fourier optics* Phys. Rev. B 79, 165116 (2009).

Chapter 2
Coupling of Spontaneous Emitters with Bloch Surface Waves

Since the pioneer work of Purcell [1], it has been clarified that the emitting properties of a light source can be strongly modified by the photonic environment surrounding it. According to the *Fermi's golden rule'* applied to the decay properties of an emitter, both the emission rate and the direction of emission are affected by the Local Density of Optical States (LDOS) [2]. In analogy with the density of electronic states, the LDOS counts for the available electromagnetic modes in which photons can be radiated from a specific location.

In proximity with a Photonic Crystal (PC) structure the radiative properties of an emitter are modified according to the photonic band structure [3] of the PC. Because of Bragg's diffraction causing constructive and destructive interference along certain directions, the far-field radiation pattern turns out to be a strongly anisotropic pattern [4]. In the case of a 3DPC exhibiting a complete band gap for a given range of frequencies, the radiative decay can be completely inhibited [5].

The high directionality achieved by employing photonic crystals has found application in different fields such as biosensing [6] or quantum information [7]. As already mentioned in the first chapter, in a phenomenological picture BSW can be seen as the photonic counterpart of SPRs. A well-known effect occurring when an emitter lies in close proximity to a metallic surface is the so called Surface Plasmon Coupled Emission (SPCE) [8]. In this case a significant portion of light radiated couples to the SPPs. The emission pattern is therefore modified according to the dispersion relation of the SPPs.

As a consequence of an enhancement of the LDOS occurring at the interface of properly designed 1DPC, an effect similar to SPCE occurs when BSW are considered [9]. In this chapter, I will show that BSW-Coupled Emission (BSW-CE) can be exploited to obtain a highly directional fluorescent emission combining the advantages of using a surface mode instead of a buried photonic mode together with the intrinsic weak ohmic losses typical of dielectric materials.

The enhanced LDOS at the truncation interface of the multilayer modifies the emitting properties of a light source thereon located [10]. The elementary cell of the design proposed here is composed by a high index layer of Ta_2O_5 ($n_h = 2.08$ at $\lambda = 600\,nm$) and a low index layer of SiO_2 ($n_l = 1.45$ at $\lambda = 600\,nm$). The thicknesses

© Springer International Publishing AG 2017
A. Angelini, *Photon Management Assisted by Surface Waves on Photonic Crystals*, PoliTO Springer Series, DOI 10.1007/978-3-319-50134-5_2

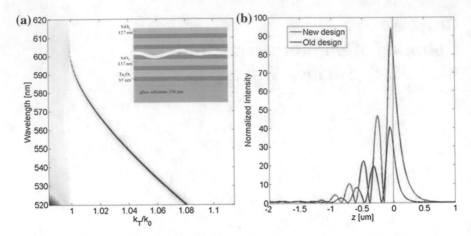

Fig. 2.1 a Calculated TE reflectance map for the multilayer design sketched in the *inset*. **b** Calculated electric field distribution of the BSW along the multilayer cross-section. The *black line* indicates the truncation interface position. The intensity has been normalized with respect to the propagating radiation in the glass. The *red line* is associated to the BSW sustained by the design in (**a**), while the *blue* one is associated to the multilayer discussed in Chap. 1

are 95 nm and 137 nm respectively, whilst the last layer of $Si O_2$ is 127 nm thick. We fabricated different samples with the same elementary cell and different number of layers, namely 8, 12 and 20 layers). The reflectance map of a multilayer with the same elementary cell and 12 layers (sketched in the inset) is reported in Fig. 2.1a. Figure 2.1b shows a comparison among the electric field distributions of the BSW modes associated to the multilayer design described in the previous chapter and the new one. As can be observed, the new design exhibits a higher enhancement of the LDOS at the truncation interface of the 1DPC, showing that the photonic properties of the sample can be tailored.

In order to evaluate the near-field coupling we modeled the field radiated by a point-like source (in the 2D model represented by an out-of-plane line of current) in a FEM model (Fig. 2.2a).

The modeling domain is a vertical cross-section of the multilayer and the emitter is placed 5 nm above the surface. The entire domain is surrounded by Perfectly Matched Layers (not shown in figure) so that boundary reflections are avoided and the domain resembles an open domain.

In order to have an effective coupling, a polarization matching between the radiated field and the BSW mode is required. The BSW is TE polarized, meaning that there are 2 possible orientations of the electric field, both matching the polarization of the BSW. We chose to have the electric field out of the plane of the cross section, in order to have the Poynting vector lying in the cross-section plane. The wavelength of the emitted radiation has been swept in the range from 560 to 600 nm (the image in Fig. 2.2a is taken at $\lambda = 580$ nm).

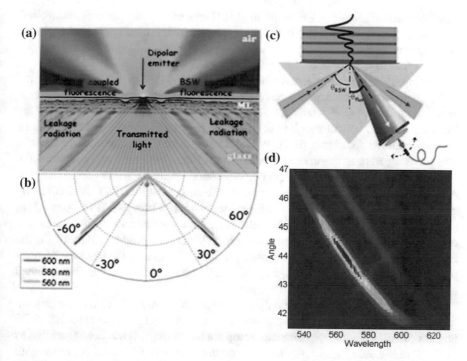

Fig. 2.2 a *Cross sectional view* if the calculated near-field distribution of the electric field radiated by a point like source. **b** Calculated far-field radiation pattern corresponding to the near-field distribution in (**a**) for different wavelengths. **c** Sketch of the experimental setup in the reverse Kretschmann configuration. Adapted from [10] **d** Experimental fluorescence intensity map function of (λ, θ). Adapted from [10]

The modeling result shows a cone of light radiated in air and one directly transmitted through the multilayer. Beside this, part of the energy couples to the BSW mode and propagates parallel to the truncation interface. The black continuous lines indicate the power flow. As long as the BSW-CE propagates at the multilayer interface, part of the energy tunnels through the multilayer and leaks into the substrate. According to the momentum conservation law, the leakage radiation propagates into the substrate at a specific angle determined by the BSW dispersion relation given by to the following relation:

$$k_{BSW} = k_0 n_{sub} \sin(\theta_{BSW}) \tag{2.1}$$

where k_{BSW} is the BSW wavevector, k_0 is the free-space wavevector, n_{sub} is the substrate refractive index (in our case $n_{sub} = 1.5$) and θ_{BSW} is the leaking angle.

Since BSW are characterized by a well-defined dispersion relation, different wavelengths will couple to BSW with a different wavevector, and will be therefore outcoupled at different angles. The far-field radiation pattern at different wavelengths has been computed (Fig. 2.2b) from the near-field distribution thanks to an algorithm based on the surface equivalent theorem [11]. Three exemplary patterns at three different wavelengths are reported in the polar plot, where it is clearly visible a variation of the output angle. In detail, longer wavelengths correspond to smaller wavevector and therefore to smaller output angles.

By employing a reverse Kretschmann configuration it is possible to recover the BSW dispersion relation in fluorescence. In Fig. 2.2c is reported a sketch of the experimental setup (for details see for example *Ballarini et al., Appl. Phys. Lett. 99, 043302*). The optical coupling between the prism and the substrate hosting the multilayer allows for extract the BSW-CE. The collection arm consists of a low NA lens focusing light at the inlet of a multi-mode fiber connected to a spectrometer. Such configuration provides high angular resolution combined with spectral analysis. An homogeneous fluorescent layer covering the surface of the 1DPC is obtained by incubating a solution of protein A labeled with Alexa Fluor 546 (PtA Alexa 546) (1 μg/ml) for 20 mins and rinsing it with PBS solution in order to remove the excess of protein. The measured map reported in Fig. 2.2d shows a bright band following the BSW dispersion relation that corresponds to the BSW-CE. In the inset an intensity profile extracted along the yellow band shows that a specific wavelength is radiated at a well defined angle with low divergence.

As an alternative way, leakage radiation can be efficiently collected by employing an high numerical aperture oil immersion objective. In order to evaluate the BSW-CE angular distribution, the experimental setup shown in Fig. 1.6 has been modified as depicted in Fig. 2.3a. Briefly, fluorescence on the surface of the 1DPC is excited with a collimated laser beam (Nd:Yag, $\lambda = 532$ nm) and imaged on the RGB CMOS camera after filtering out the excitation light with an edge filter (Semrock 532 nm RazorEdge). The adjustable tube lens in front of the CMOS camera allows to image both the direct plane and the BFP of the oil immersion objective [12]. Figure 2.3b shows an example of BFP imaged when fluorescence is excited on the flat surface of a 1DPC. The outer and inner dashed white circles indicate respectively the NA of the objective and the light line. The fluorescence ring lying right beyond the light line corresponds to the projection of the cone of emission resulting from the leakage of the BSW-CF, schematically depicted in Fig. 2.3a. The RGB image allows to qualitatively evaluate the BSW dispersion.

The fluorescence ring is characterized by an external green area that turns into red light approaching the light line. A cross-section of R and G channels along the yellow dashed line allows to better evaluate the dispersive behavior of BSW-CE and compare the experimental results to the theoretical predicted output angles. The normalized transverse wavevector ($k_T/k_0 = n_{eff}$) associated to shorter wavelengths (green) of the spectrum has its maximum at 1.09, corresponding to an output angle of 46.6°, while the peak associated to longer wavelengths (yellow-red) is centered at $n_{eff} = 1.05$ (44.4°). Such values can be compared to the angles of emission calculated for $\lambda_G = 540$ nm and $\lambda_R = 600$ nm, that are respectively 46.5° and 44.5°.

Due to the polarization selectivity of the BSW mode, only dipoles whose momentum is oriented parallel with the surface will efficiently couple with the BSW. As a consequence, BSW-CE will exhibit a well defined polarization as well, and it will be

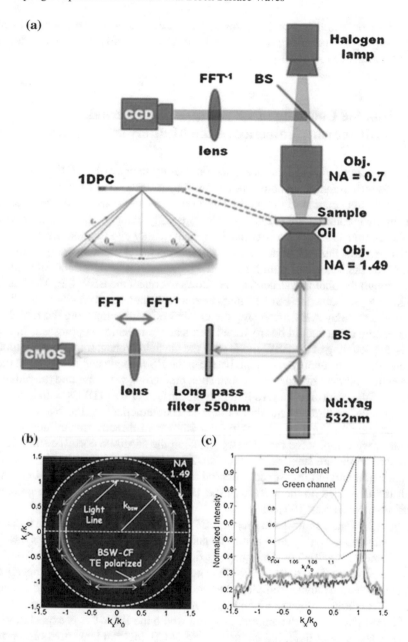

Fig. 2.3 **a** Schematic view of the customized fluorescence leakage radiation microscope. **b** Example of real color fluorescence BFP obtained when fluorescence is excited on the flat surface of a 1DPC. The external *white dashed ring* indicates the NA of the oil immersion objective, while the internal one signs the air/glass light line. The light polarization state is indicated by the *white arows*. An intensity cross-section is reported in (**c**), where the normalized intensities collected by the R and G channels of the CMOS camera are reported. In the *inset* a detail about the angular position of the two peaks

azimuthally polarized (see white arrows) in the specific case of TE polarized BSW (conversely, SPCE occurring on a flat metallic film is radially polarized, reflecting the TM nature of SPPs [13]).

2.1   Tunable Coupling of BSWs with the External Environment: Influence of the Multilayer Thickness

Several factors may contribute to reduce the quality factor of BSW. Defects on the surface as well as inhomogeneities of the multilayer stack may give rise to scattering of the radiation coupled to the mode. Ohmic losses can be also taken into account although their contribution is rather weak, especially in the visible range. Provided that the multilayer exhibits a reasonable optical quality and the materials are characterized by low intrinsic ohmic losses, the main channel draining energy out from the mode is the leakage of light through the multilayer. It is indeed the tunneling of light through the photonic structure that allows to collect the BSW-CE. The leakage of light into the substrate can be interpreted as a tunneling event due to the finiteness of the multilayer. By increasing the number of pairs composing the multilayer the tunneling event should be frustrated and leakage radiation suppressed. In other words, the coupling of the BSW mode with the far-field propagating substrate modes is reduced, and the energy exchange between the BSW mode and the external environment is inhibited. In order to confirm such prediction, we computed the radiation patterns of the point-like sources lying on the surface of two 1DPCs with same elementary cells but different number of layers. The numerical results are reported in the polar plots in Fig. 2.4a. By taking the amount of directly transmitted light as reference, we can observe that the two peaks in the substrate corresponding to the leakage radiation are suppressed when the *10 pairs* design is considered, meaning that the major part of fluorescence coupled to BSW has not been radiated in far field and it is still guided on the surface of the 1DPC at the boundaries of the modeling domain which is about 100 μm.

Figure 2.4b and c show the fluorescent BFP collected when fluorescence is excited on the flat surface of a *4 pairs* (b) and 10 pairs (c) 1DPCs (the two samples have the same elementary cell). The first image is dominated by the bright ring beyond the light line corresponding to the BSW-CE. The second one appears brighter because of a higher excitation intensity employed to put in evidence the photonic band structure in the BFP, confirmed by the much higher intensity of the directly transmitted light within the light line. At the angular position where the BSW-CE is expected to be, there is a shortage of light, indicating that BSW-CE does not leak anymore into the substrate and it is therefore no more collectable in far field.

In order to have an easier comparison, the images reported in (b) and (c) have been merged in a single image. The angular position of the BSW is characterized by a peak of fluorescence in the *4 pairs* design, while at the same position a shortage of light is observable when the *10 pairs* design is considered. Therefore, in the low leakage

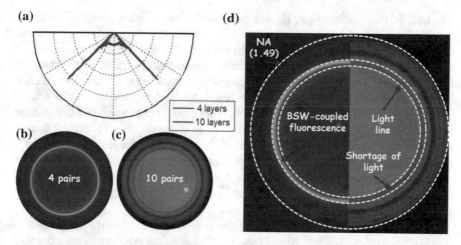

Fig. 2.4 Calculated far field radiation pattern for two multilayers with same design and four pairs of layers (*blue line*) and ten pairs of layers (*red line*). **b, c** Experimental fluorescence measure showing the BFP collected when fluorescence is excited on the flat area of a four pairs **b** and ten pairs **c** 1DPC. The two images are merged in (**d**) for easier comparison

design, we can assume that BSW-CE is forced to be confined at the surface and the main factor reducing the propagation length are the surface defects and ohmic losses of the layers, that are determined by the overall quality of the fabrication process.

The possibility of controlling the leakage radiation and eventually suppress it gives an interesting advantage in this framework: the interaction of fluorescence coupled to the surface modes with distributed surface structures can be indeed enhanced if competitive radiative decay channels are inhibited.

2.2 Directional Extraction of Luminescence Assisted by Surface Relief Gratings

The ability of controlling the directivity of spontaneous emission is an objective of outstanding relevance in a variety of fields dealing with fluorescent emitters, such as biosensing when labeled markers are employed [4] or quantum information applications based on transmission of single photons emitted by single photon sources [14]. The enhanced directionality of spontaneous emission would allow to enhance the collection efficiency of fluorescence by employing low numerical aperture systems that can operate also from the air side (and not necessarily from the substrate as in the case of oil-immersion optics).

A grating coupler can also be exploited to efficiently extract the BSW-CE into a collimated beam. Such effect has been widely reported when SPCE on a periodically corrugated metallic film is considered [15, 16]. Although an effective beam generation assisted by Surface Plasmons (SPs) has been reported by several groups, the performances of such systems are often limited by the typical decay lengths of

SPs that, in the visible range, do not exceed few tens of microns [17]. Moreover the directionality of emission is intrinsically limited by the broadness of SPs resonances, mainly due to ohmic losses. For the above mentioned reasons, we expect that longer decay length as well as a narrower resonances provided by the photonic crystal platform would allow for higher directionality of the extracted fluorescence.

The directional extraction of radiation coupled to resonant modes may be critical in some applications. The use of immersion optics or prisms to extract the leakage radiation might be limiting or unfeasible. As an example, the prism may impose a strong constrain when an image of the surface based on the leakage radiation is required [18], while the alignment of immersion optics may be not trivial. The use of relief gratings patterned on the surface of the 1DPC should allow to diffract energy coupled to BSW into free-space radiation by adding a specific momentum given by the grating periodicity, according to the Bragg's law.

In the following, we will consider a linear grating patterned on the surface of two 1DPC samples. The sequence of the multilayer stacks considered here is $glass$-$N[Ta_2O_5$ (95 nm)-$SiO_2(137$ nm)]-Ta_2O_5 (95 nm)-SiO_2 (127 nm)-air, where N can be 4 or 10 (see Chap. 1). Given the dispersion relation of the BSW, the Bragg's law can be used to determine the grating periodicity needed to normally diffract BSW-coupled emission, at least to a first approximation. It has to be noticed indeed that the fabrication of the grating modifies the dispersion relation of the BSW, because of the dielectric loading effect (Sect. 1.2).

In order to optimize the grating parameters (i.e. period and corrugation depth), the FEM model proposed in Chap. 1 (emitter on flat surface, Fig. 2.5a, b) has been modified by introducing a surface corrugation (Fig. 2.5c, d). In the model the diffractive structure has the same refractive index of SiO_2 ($n_{SiO_2} = 1.45$ at $\lambda = 600$ nm). The grating parameters were optimized in order to have a normally diffracted beam at $\lambda = 570$ nm, i.e. the peak emission wavelength of PtA Alexa 546 which is the fluorescent probe used in the experimental observations. The parametric study over the diffraction efficiency as a function of the corrugation height indicates that the maximum efficiency occurs when the grating height is about 80 nm (calculations not shown), while the periodicity needed to normally diffract BSW-CE at $\lambda = 570$ nm is $\Lambda_g = 520$ nm. The near field distribution (Fig. 2.5d) shows that fluorescence radiated couples with the surface modes and undergoes scattering events as long as it propagates parallel to the truncation interface. By enlarging the simulation domain (Fig. 2.5c), far from the surface it is visible the formation of a normal beam due to the coherent superposition of the scattered light. By computing the far-field pattern (Fig. 2.5d), a pronounced normal beam is generated both in air and in the substrate, while the two lateral beams associated to leakage radiation of BSW-CF on a flat surface (number of layer pairs N = 4) (here reported for easy comparison) are no more visible. The far-field radiation pattern shown in the figure refers to the cumulative radiation patter obtained by integrating over the emission wavelength range of 550–600 nm and it predicts a divergence of the overall beam of $\approx 5°$ in air.

In order to observe the angular distribution of fluorescence, the leakage radiation microscope described in the first chapter is used (Fig. 1.6). The diffractive structure considered is the one schematically shown in Fig. 2.6a.

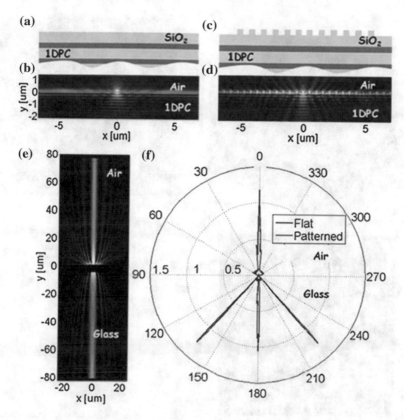

Fig. 2.5 Cross sectional sketch of flat (**a**) and patterned (**c**) multilayer; calculated near field intensity distribution of light radiated by an emitter close to the surface of a flat (**b**) and patterned (**d**) multilayer (N = 10); **e** expanded view of the intensity distribution of BSW-coupled diffracted fluorescence; **f** angular pattern of emitted light in the case of flat and corrugated 1DPCs, highlighting the BSW assisted beaming effect. Adapted from [22]

Figure 2.6b shows the fluorescence image of the BFP collected on the N = 4 pairs design. In addition to the bright circle corresponding to the leakage of BSW-coupled fluorescence naturally occurring on the flat multilayer, the BFP fluorescence image reveals two bright arcs superposing in the center. The two additional arcs correspond to the diffracted BSW-CE, and the superposition in the center of the first diffracted orders confirms that the grating momentum matches the BSW wavevector. In this case, because of the particular orientation of the linear grating, diffraction only affects the k_x direction of propagation, through the corresponding wavevector components:

$$k_x = (2\pi/\lambda)n_{sub}sin(\theta_x) \tag{2.2}$$

for each wavelength λ.

The sample with lower number of layers is considered here in order to put in evidence the correspondence of the BSW-CE and the diffracted light. It has to be

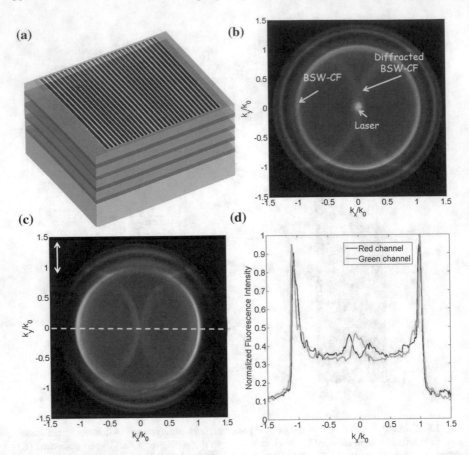

Fig. 2.6 **a** Schematic view of the linear grating patterned on top of the 1DPC. The *top layer* is a SEM image. Adapted from [23] **b** Fluorescence BFP image obtained by collecting fluorescence coming from the patterned area. The two diffracted branches superimposing in the center confirm the BSW-coupled fluorescence diffraction. The *bright green spot* in the center is the residual laser radiation. **c** Same of (**b**) collected with a polarization filter (oriented according to the *white arrow*) along the collection path that puts in evidence the TE polarization of both the BSW-coupled fluorescence and the diffracted BSW-CE. **d** Intensity profile extracted along the *red dashed line* in (**b**). that evidences the normal beam due to the grating effect

noticed indeed that the two branches crossing in the center are the replicas of the bright ring corresponding to BSW-CE leakage radiation shifted in the *x* direction of an amount equal to the grating momentum.

In order to put in evidence the polarization dependence of the BSW coupling, a polarization filter has been added along the collection path (Fig. 2.6c). The image reveals the TE nature of BSWs and confirms that the diffracted light maintains the polarization state (along the white arrow), confirming that it is the BSW-CE that is diffracted.

We can extract an intensity profile for both the red (R) and green (G) channels of the CMOS camera (Fig. 2.6d). The profile is taken along the $k_y = 0$ (dashed) line in Fig. 2.6c, and the appearance of a central peak can be observed. The R channel has its maximum of sensitivity in the spectral range 580–620 nm, that is the same spectral range of the far-field pattern calculations in Fig. 2.5f. We can therefore apply the Bragg's equation for a $\Lambda_g = 520$ nm grating. For a given wavelength λ, we find that $k_x^{-1} = k_x^{BSW} - 2\pi/\Lambda_g$ results in an overall angular range $\theta_x \approx 6°$ associated with the -1 diffraction order of BSW-coupled fluorescence. The chromatic dispersion of the diffracted branches reflects the dispersion of BSWs. The experimental findings well match theoretical predictions obtained with the simple two-dimensional finite element numerical model described above.

2.3 Enhanced Fluorescent Signal in Bio-Sensing Experiments

So far we have considered the sample with the lower number of layers in order to qualitatively show that BSW-CE can be normally diffracted by means of surface relief gratings. In the attempt to take advantage of this effect in an application, a quantification of the enhancement obtained for a given numerical aperture of the collection system is required.

To this aim, the experimental setup has been modified according to the scheme presented in Fig. 2.7. A linearly polarized 532 nm laser beam (Nd:Yag 10 mW) is expanded by means of a 0.1 NA objective and a movable lens with focal length 10 mm. The divergence of the beam can be adjusted by moving the lens. Since the grating is fabricated to normally diffract fluorescence, a small divergence of the laser beam is required in order to couple the incident radiation with the BSW and ensure a resonant excitation of fluorescence. The beam passes through an Half Wave Plate able to rotate the polarization of the incident beam and then impinges on the sample. The fluorescence excited on the surface of the 1DPC is collected by a low numerical aperture objective (Nikon 5x NA = 0.1) and spectrally filtered by an edge filter (FEL 550, Thorlabs) in order to cut off the excitation light. Fluorescence is then imaged on a CCD camera (Apogent Ascent a694) via a lens so that a real image of the sample surface is produced.

The sample is schematically depicted in Fig. 2.8a (image not in scale): a 1DPC composed by six pairs of high and low refractive index layers (same elementary cell described in the previous section) is patterned with a linear surface relief grating (period $\Lambda_g = 495$ nm, height $h = 80$ nm). The patterned area is composed by four groups of six squares of 250 μm $\times$ 250 μm size. The gratings are clearly visible by naked eyes under white light illumination (Fig. 2.8b, c). An homogeneous layer of a fluorescent probes is obtained by functionalizing the surface and binding thereon the labeled molecules. Different models have been tested on the same sample, demonstrating the versatility of the approach. More details about the biological models and

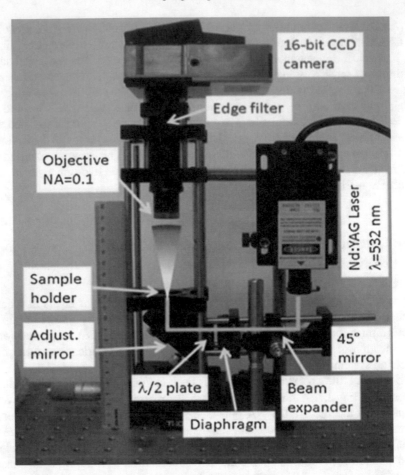

Fig. 2.7 Picture of the fluorescence microscope used in the sensing experiments. Adapted from [20]

the related protocols for functionalize the photonic crystal surface can be found in *F. Frascella et al., Analyst 140, 5459* and *S. Ricciardi et al., Sens and Act B 215, 225.* The sample itself can be eventually equipped with a microfluidic chip for on-chip functionalization [19].

Since the signal amplification is not related to the specific molecule, we can choose a generic model to quantify the enhancement factor. To this aim we consider the c-DNA mi-RNA-16 (labeled) recognition reported in [20]. We chose a specific mi-RNA concentration (50 nM) well below the saturation level. The resonant coupling of the laser beam with the BSW mode can be easily switched on and off by rotating the Half-Wave Plate. The couping occurs indeed only when the laser polarization is perpendicular to the grating momentum. Figure 2.9a, b correspond to the fluorescent images obtained without resonant excitation and with resonant excita-

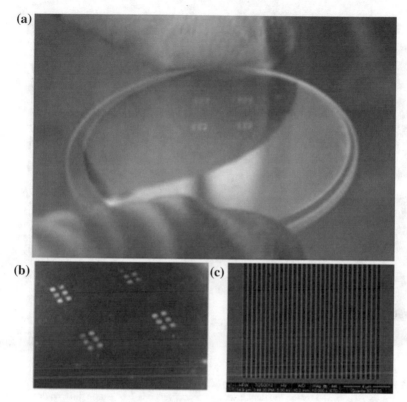

Fig. 2.8 Schematic view of the patterned 1DPC. **b** Picture of the four groups of linear gratings under white light illumination. **c** Image of the photonic crystal sample employed in the sensing experiments. Adapted from [20]

tion respectively. Since the fluorescent probes are homogeneously distributed on the sample surface and the laser spot is much larger than a single grating pad, the signal coming from outside the patterned areas can be taken as internal reference. By taking the intensity profiles along the two yellow dashed lines in Fig. 2.9a, b, we can compare the two signals (Fig. 2.9c). The purple line corresponds to the fluorescence intensity in off-resonance condition. In this case the enhancement factor observed is about 3 and it is only due to the angular redistribution of light. The black line is corresponds to the case of resonant excitation and an enhancement factor of about 60 is obtained, meaning that the resonant excitation contributes to the overall enhancement with a factor of roughly 20.

By incubating solutions at different concentrations of mi-RNA 16 it is possible to build a titration curve (Fig. 2.10). To this aim, we considered concentrations ranging from 0.5 nM to 2.5 μM. Each concentration has been incubated on a group of six grating pads, in order to perform a statistical analysis. The error bars result from the intensity fluctuations observed on different gratings. Both the two sets of experimental data are well fitted by a sigmoidal curve, as reported in literature [21].

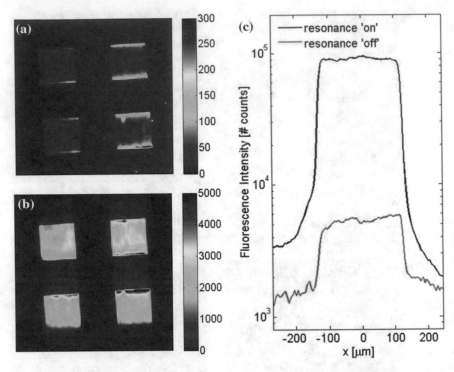

Fig. 2.9 Fluorescence image (*in false colors*) of a sub-group of 4 gratings either non-resonantly (**a**) or resonantly (**b**) excited by the incident laser beam (incidence is 1.6°) upon polarization rotation; **c** comparison of intensity profiles across one grating (*yellow dashed line* in (**a, b**)) under on-resonance and off-resonance conditions. Adapted from [20]

The two background levels have been obtained by measuring the fluorescence intensity before mi-RNA 16 incubation both inside the patterned area and on the surrounding flat surface. The two distinct level of background are due to the enhancement affecting also the autofluorescence of c-DNA and silica. The minimum detectable signal is set as 3σ, where σ is obtained by computing the root mean square of the background intensity. The two curves demonstrate that the minimum concentration detectable on the flat surface is about 50 nM, whilst on the patterned area a concentration as low as 0.5 nM is still distinguishable from the background. As a result, the limit of detection is reduced of roughly 50 in concentration, and a mean enhancement factor of 60 is maintained over the entire range of concentrations.

The approach has been successfully tested also with an antigen-antibody model [22], demonstrating the versatility of the approach combined with the simplicity of the detection scheme, that can be easily implemented in a commercial wide field fluorescence microscope.

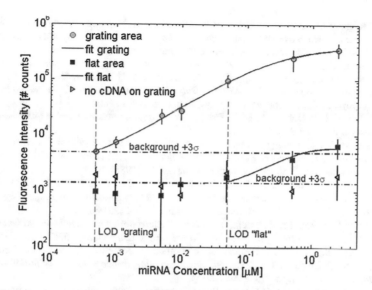

Fig. 2.10 Fluorescence titration curves collected for different miRNA-16 concentrations, inside and outside the gratings. The reported concentration refers to the miRNA-16 molarity in the incubating solution. Adapted from [20]

References

1. E. M. Purcell, Phys. Rev. 69, 681, (1946).
2. R. S. Meltzer, S. P. Feofilov, B. Tissue and H. B. Yuan, *Dependence of fluorescence lifetimes of* $Y_2O_3 : Eu_3^+$ *nanoparticles on the surrounding medium*, Phys. Rev. B 60, R14012(R), (1999).
3. M. Megens, J. E. G. J. Wijnhoven, A. Lagendijk and W. L. Vos, *Fluorescence lifetimes and linewidths of dye in photonic crystals* Phys. Rev. A 59, 4727, (1999).
4. N. Ganesh, W. Zhang, P. C. Mathias, E. Chow, J. A. N. T. Soares, V. Malyarchuk, A. D. Smith and B. T. Cunningham, *Enhanced fluorescence emission from quantum dots on a photonic crystal surface*, Nat. Nanotech, 2, 515–520, (2007).
5. P. Lohdal, A. F. van Driel, I. S. Nikolaev, A. Irman, K. Overgaag, D. Vanmaekelbergh and W. L. Vos, *Controlling the dynamics of spontaneous emission from quantum dots by photonic crystals.*, Nature 430, 654–657, (2004).
6. I. D. Block, L. L. Chan, B. T. Cunningham, *Photonic crystal optical biosensor incorporating structured low-index porous dielectric*, Sens. and Act. B 120, 187–193, (2006).
7. W.-H. Chang, W.-Y. Chen, H.-S. Chang, T.-P. Hsieh, J.-I. Chyi and T. Hsu, *Efficient Single-Photon Sources Based on Low-Density Quantum Dots in Photonic-Crystal Nanocavities*, Phys. Rev. Lett. 96, 117401, (2006).
8. J. R. Lakowicz, *Radiative decay engineering 3. Surface plasmon-coupled directional emission*, An. Biochem. 324 (2), 153–169, (2004).
9. R. Badugu, K. Nowazcyk, E. Descrovi, J. R. Lakowicz, *Radiative decay engineering 6: Fluorescence on one-dimensional photonic crystals.*, An. Biochem. 442 (1), 83–96, (2013).
10. M. Ballarini, F. Frascella, N. De Leo, S. Ricciardi, P. Rivolo, P. Mandracci, E. Enrico, F. Giorgis, F. Michelotti, and E. Descrovi, *A polymer-based functional pattern on one-dimensional photonic crystals for photon sorting of fluorescence radiation* Opt. Express 20, 6703 (2012).
11. Taflove A. and Hagness S. C. *Computational Electrodynamic: The Finite Difference Time Domain Method* 3rd edn (Boston MA: Artech House) p. 329, (2005).

12. C. J. Regan, O. Thiabgoh, L. Grave de Peralta, and A.A. Bernussi, *Probing photonic Bloch wavefunctions with plasmon-coupled leakage radiation*, Opt. Expr. 20 (8), 8658–8666, (2012).
13. S.P. Frisbie, C.J. Regan, A. Krishnan, C. Chesnutt, J. Ajimo, A.A. Bernussi and L. Grave de Peralta, *Characterization of polarization states of surface plasmon polariton modes by Fourier-plane leakage microscopy*, Opt. Comm. 283 (24), 5255–5260, (2010).
14. R. Esteban, T. V. Teperik, and J. J. Greffet, *Optical Patch Antennas for Single Photon Emission Using Surface Plasmon Resonances*, Phys. Rev. Lett. 104, 026802, (2010).
15. Y. C. Jun, K. C.Y. Huang and M. L. Brongersma*Plasmonic beaming and active control over fluorescent emission*, Nat. Comm. 2, 283, (2011).
16. H. Li, S. Xu, Y. Gu, H. Wang, R. Ma, J. R. Lombardi and W. Xu, *Active Plasmonic Nanoantennas for Controlling Fluorescence Beams*, J. Phys. Chem. C 117, 19154–19159, (2013).
17. H. Raether, *Surface Plasmons*, Springer-Verlag, Berlin (1988).
18. E. Descrovi, D. Morrone, A. Angelini, F. Frascella, S. Ricciardi, P. Rivolo, N. De Leo, L. Boarino, P. Munzert, F. Michelotti and F. Giorgis, *Fluorescence imaging assisted by surface modes on dielectric multilayers*, Eur. Phys. Journ., 68, 1–4 (2014).
19. A. Lamberti, A. Angelini, S. Ricciardi and F. Frascella, *A flow-through holed PDMS membrane as a reusable microarray spotter for biomedical assays*, Lab on Chip 15, 67–71, (2015).
20. F. Frascella, S. Ricciardi, L. Pasquardini, C. Potrich, A. Angelini, A. Chiado', C. Pederzolli, N. De Leo, P. Rivolo and E. Descrovi, *Enhanced fluorescence detection of miRNA-16 on a photonic crystal*, Analyst 140, 5459, (2015).
21. B. T. Cunningham and R. C. Zangar, *Photonic crystal enhanced fluorescence for early breast cancer biomarker detection* J. Biophotonics 5, 617 (2012).
22. S. Ricciardi, F. Frascella, A. Angelini, A. Lamberti, P. Munzert, L. Boarino, R. Rizzo, A. Tommasi and E. Descrovi, *Optofluidic chip for surface wave-based fluorescence sensing*, Sens. and Act. B 215, 225–230, (2015).
23. A. Angelini, E. Enrico, N. De Leo, P. Munzert, L. Boarino, F. Michelotti, F. Giorgis and E. Descrovi, Fluorescence diraction assisted by Bloch surface waves on a one-dimensional photonic crystal, New J. Phys. 15, 073002, (2013).

Chapter 3
Ring Antennae: Resonant Focusing and Collimated Extraction of Luminescence

In the previous chapter it has been demonstrated that a periodic linear corrugation patterned on the surface of a 1DPC can extract the BSW-CE. Due to the spatial invariance of the surface structure, such configuration does not require a specific localization of the emitter. When a localized ensemble of emitters randomly oriented is considered, the main drawback of such simple approach is that the extraction efficiency is not maximized. In this case the natural coupling of the emitted radiation with BSW occurs along all the in-plane directions, whilst the grating works properly only along a specific direction given by the grating momentum orientation. In the framework of Surface Plasmon Polaritons it ha been demonstrated that a diffractive structure exhibiting circular symmetry can produce a bidimensionally collimated beam when either a scattering center like a subwavelength hole or spontaneous emitters are located in the center of symmetry of the structure [1–4]. We therefore decided to apply the same approach to extract BSW-CE coming from localized sources [5].

The structure that we are going to analyze is the one schematically shown in Fig. 3.1. The elementary cell of the multilayer stack is the same described in the previous chapter and in the following we will consider two layouts with the same elementary cell: the first one is composed by 16 layers, whilst the second one is composed by 22 layers. In the latter case the leakage radiation is almost completely suppressed and the interaction of BSW-CE with the diffractive structure is therefore maximized. The circular grating is etched on the top layer and surrounds a flat inner spacer of few microns in diameter. The total thickness of the grating is about 80 nm and the corrugation has a spatial period $\Lambda_g = 520$ nm. The grating vector is radially oriented with respect to the grating center, and has a module $G = 2\pi/\Lambda_g = 12.08 \, \mu m^{-1}$ that is very close to the BSW wave-vector in the wavelength range between 532 and 590 nm wavelength.

© Springer International Publishing AG 2017
A. Angelini, *Photon Management Assisted by Surface Waves on Photonic Crystals*, PoliTO Springer Series, DOI 10.1007/978-3-319-50134-5_3

(a)

(b)

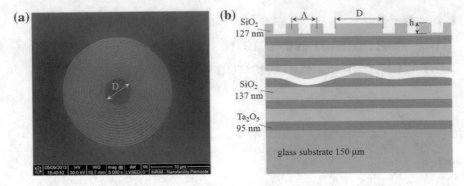

Fig. 3.1 **a** SEM image of the concentric ring structure patterned on the surface of a 1DPC. The structure is etched in the last SiO_2 layer. **b** Schematic cross-sectional view of the diffractive structure engraved in the last layer of the photonic crystal. Copyright Nature Publishing Group. Reprinted from [5]

3.1 In-Plane Focusing by Concentric Corrugations

In Chap. 1 we have seen that ultra-thin refractive structures can be employed to focus BSW-coupled radiation in a small spot. As an alternative strategy to enhance the electromagnetic field in a localized area on the surface of the multilayer, diffractive structures can be considered [5]. Differently from the refractive structures presented in the previous section where the coupling of far field radiation and the focusing effect of the fabricated polymeric structures where independent events, here the structure itself allows for coupling the incident far-field radiation to the BSW mode, while the peculiar symmetry of the structure provides the focusing effect.

The experimental setup employed to perform the following experiments is the one schematically shown in Fig. 1.17, with an additional feature: a beam blocker placed in the center of the BFP of the collection objective (Fig. 3.2) blocks the

Fig. 3.2 Schematic view of the beam blocker placed in the BFP of the collecting objective to filter out the directly transmitted light

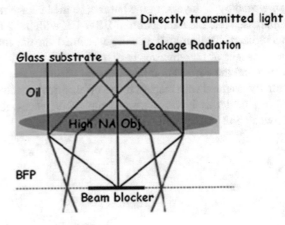

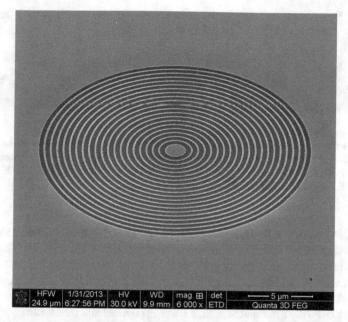

Fig. 3.3 SEM image of the concentric ring diffractive structure employed to couple far-field radiation to BSW and localize it in the center of the ring array. The concentric grating is patterned on the surface of the 1DPC by EBL

directly transmitted light. By this way, the image formed in the CCD camera is mainly produced by the collected leakage radiation, allowing to image the spatial distribution of radiation coupled to BSW on the surface.

The sample structure is reported in Fig. 3.3. Very briefly, the surface of the 1DPC is patterned with an array of concentric rings surrounding an inner flat spacer. The circular grating, fabricated by Electron Beam Lithography on PMMA, is etched on the 1DPC top layer by a Reactive Ion Etching step followed by lift-off in acetone to remove the resist. The concentric ring pattern is arranged with a spatial period $\Lambda_g = 480$ nm, and the total etching depth is about 100 nm. The inner spacer diameter can be 5 or 8 μm. The grating vector, radially oriented with respect to the grating center, has a module $G = 2\pi/\Lambda_G = 13.09\,\mu\mathrm{m}^{-1}$, that is very close to the BSW wavevector at $\lambda = 532$ nm. When the laser radiation (CW doubled Nd:Yag at 532 nm) radiation impinges on the circular grating from air, it undergoes diffraction according to the Bragg's law. If we consider the wavevector components parallel to the truncation interface of the multilayer, the first-order (+1) diffracted beam has a wavevector component k_T^{+1} given by $k_T^{+1} = k_T^0 + G$, where k_T^0 is the wavevector component of the incident radiation. The coupling, i.e. the energy transfer, between the incident far field radiation and the BSW mode occurs when k_T^{+1} matches the BSW wavevector.

Due to the symmetry of the grating considered here, we can expect a focusing effect of the structure itself on the radiation coupled to the BSW mode. The radius of curvature of the grating should indeed direct the coupled light toward the center of the

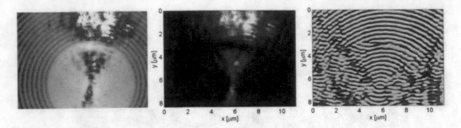

Fig. 3.4 In sequence: white light image of the patterned surface superposed to the intensity distribution obtained by collecting only the leakage radiation. Amplitude and phase distributions of the optical fields collected at the surface with the same field of view of the first image. Copyright Nature Publishing Group. Reprinted from [5]

structure, in analogy to surface plasmons [6, 7]. In order to experimentally observe such focusing effect, we locally illuminated a portion of the grating. The leakage radiation associated to BSW is collected with the interference leakage radiation microscope shown in Fig. 1.17, to which a beam blocker has been added, according to the working principle illustrated in Fig. 3.2.

Starting from the illuminated area, a BSW is coupled and then focused in a central region of the inner spacer surrounded by the circular grating. A superposition of a white image of the structure with the intensity distribution of light, amplitude and phase maps of the focused BSW are shown in Fig. 3.4.

More interestingly, when a collimated beam homogeneously illuminates the circular grating, the coupling to BSW can occur along all the in-plane directions, according to the orientation of the grating. A circularly symmetric BSW converging toward the center can be then produced. The illumination path can be slightly adjusted in order to modify the divergence of the incident beam, in such a way that the momentum matching between BSW and the incident radiation is achieved. The experimental results showing the amplitude and phase maps of leakage radiation collected are presented in Fig. 3.5a and b. Since the BSW is s-polarized, a linearly polarized beam can be coupled to BSW only from those regions of the rings where the polarization matching condition is satisfied in addition to the momentum matching condition.

More specifically, here the coupling occurs only if the polarization of the incident beam is perpendicular with respect to the grating momentum. In Fig. 3.5a (y-polarized incident beam), the leakage radiation associated to the focused BSW has a two-lobe symmetric distribution oriented along the x-axis. When the laser polarization is rotated, the observed pattern rotates accordingly.

The radiation coupled to the BSW mode converges toward the center of the inner spacer, in a small spot. By analyzing the cross-sectional profile of the field distribution (Fig. 3.5c), we can see a central lobe in the spacer center, whose estimated Full Width Half Height is about 200 nm. Moreover, the amplitude profile shows an interference pattern with a spatial frequency doubled as compared to the wavefront frequency appearing in the cross-sectional profile extracted from the phase map (Fig. 3.5d). The interference here occurs because of the superposition of two counter-propagating

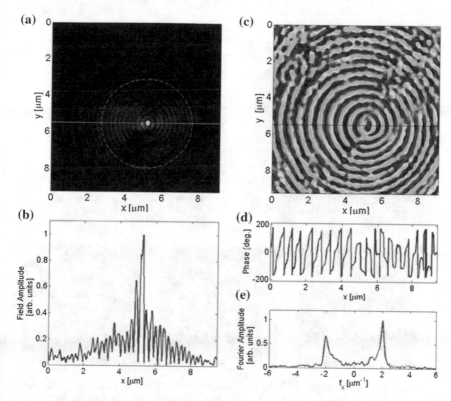

Fig. 3.5 **a** amplitude and **b** phase distribution of the leakage radiation associated to a converging laser BSW. Coupling is performed by illuminating a circular grating ($D = 5\,\mu m$, $l - 520\,nm$) on a "regular" multilayer with a linearly polarized (y-direction) laser ($\lambda = 532\,nm$). The *dashed circle* indicates the boundary on the inner spacer. **c** Cross sectional field amplitude profile along the *horizontal line* shown in (**a**). **d** Cross-sectional field phase profile and **e** its corresponding Fourier spectrum amplitude along the *horizontal line* shown in (**b**). Copyright Nature Publishing Group. Reprinted from [5]

BSWs. The sawtooth profile appearing in the phase cross-section along the horizontal line sketched in Fig. 3.5b indeed indicates the presence of two counter-propagating surface waves converging toward the grating center, where a phase singularity appears (Fig. 3.5d). The coexistence of the two counter-propagating BSWs can be better noticed by looking at the Fourier spectrum amplitude as calculated from the complex field distribution (obtained by combining the measured amplitude and phase maps) along the horizontal line (Fig. 3.5e). The Fourier spectrum amplitude exhibits two main peaks centered at spatial frequencies $f_x = 62.02\,\mu m^{-1}$, that matches the calculated BSW wavevector obtained from the BSW dispersion relation.

A less detailed but easier way to observe the focusing effect of the circular grating can be the wide field fluorescence image of the grating structure. The fluorescent signal may indeed map the near field distribution of the excitation field, provided

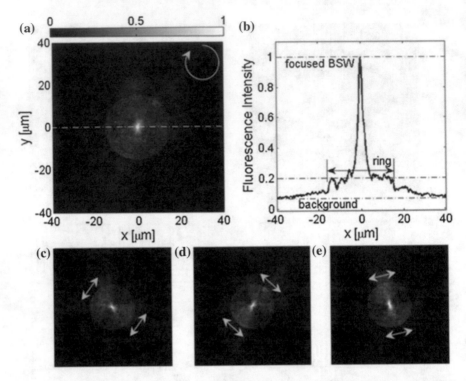

Fig. 3.6 a Fluorescence image of a laser BSW coupled and focused by a circular grating ($D =$ 8 μm, $L = 520$ nm) on "low leakage" multilayer. Illumination is a collimated, circularly polarized laser beam. **b** cross-sectional fluorescence intensity along the *dashed line* in (**a**). **c–e** same as in (**a**) with an incident laser beam linearly polarized as sketched by the *arrows* in the figures. All images are collected by means of the setup sketched in Fig. 1.17. Copyright Nature Publishing Group. Reprinted from [5]

that the excitation intensity is well below the saturation level. In order to provide an homogeneous fluorescent layer over a large area, a solution containing Alexa Fluor 546-labeled protein A was spun over the sample, thereon incubated for 20 min and then rinsed with PBS solution. Figure 3.6 shows direct plane fluorescence images of a diffractive ring structure illuminated by a collimated laser beam much larger than the structure.

The setup used for these measurements is the fluorescence microscope described in Chap. 1, Fig. 1.4. Here, it is worth to underline that the collection objective has a numerical aperture of 0.2 and operates in air, in such a way that only directly transmitted light contributes to the fluorescence image. The fluorescence image shown in Fig. 3.6a is collected when a circularly polarized beam is used for excitation. The image shows some fluorescence radiated from the surface structure surrounding a bright spot located in the grating center. Although the fluorescent probes are homogeneously distributed on the whole sample surface, the signal collected from outside the ring grating is rather weak. If we consider a normalized intensity profile taken along the horizontal dashed line drawn in Fig. 3.6a, the inhomogeneity of the intensity distribution across the structure can be better appreciated. The central peak corresponding to the bright central spot is roughly 10 times higher than the fluorescent signal coming from the outside of the ring (Fig. 3.6b). The enhancement of the fluorescent signal is attributed to the BSW coupling and focusing of the incident laser illumination. Since the inner spacer is flat and rather wide (five to ten microns), any accumulation of the organic molecules on the surface can be safely excluded. The enhancement of the excitation field is further confirmed by using a linearly polarized incident beam: in this case, the fluorescence images in Fig. 3.6c–e show a two-lobed shape, and the angular position of the lobes depends on the orientation of the laser polarization.

3.2 Beaming of Fluorescence from Localized Sources

In order to characterize the concentric ring grating behavior, we incubated a solution containing Pt A labeled with Alexa Fluor 546 (1 µg/ml) for 20 min and then rinsed it with PBS, so that the surface was covered by an homogeneous fluorescent layer. We then focused the laser beam (CW doubled Nd:Yag 532 nm) in a diffraction limited spot entirely contained within the inner flat spacer (Fig. 3.7a). In this way we can avoid laser coupling with the BSW mode and be sure that the fluorescence is excited only within the flat area.

Figure 3.7b shows a wide field fluorescence image of the sample surface collected with a 0.95 NA objective (Nikon 100x 0.95). The fluorescence image reveals a bright

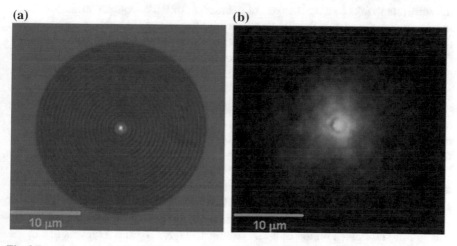

(a)

(b)

10 µm

10 µm

Fig. 3.7 **a** Reflection bright-field image of a circular grating made through a top objective (NA = 0.95) and laser focusing in the central spacer; **b** direct plane fluorescence image collected by the NA = 0.95 objective operating in air. Reprinted from [8]

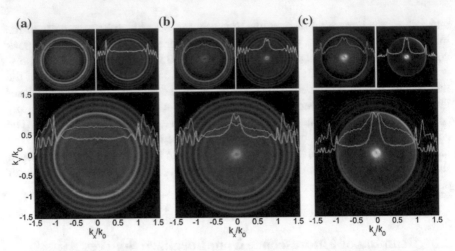

Fig. 3.8 Back focal plane images for BSW-coupled fluorescence. **a** Fluorescence BFP images collected from a planar *regular* multilayer. **b** Fluorescence BFP collected from a circular grating ($D = 5\,\mu m$, $\Lambda_g = 520\,nm$) fabricated on *regular* multilayer. **c** Fluorescence BFP collected from a circular grating ($D = 5\,\mu m$, $\Lambda_g = 520\,nm$) fabricated on *low leakage* multilayer. Illumination is a laser beam focused onto the center of the inner spacer. For each colour image, the corresponding R-channel (*red*) and G-channel (*green*) images are separately shown in a *row* on *top* of the figure. Copyright Nature Publishing Group. Reprinted from [5]

central spot from where fluorescence directly radiates in air. More interestingly, fluorescence is also detected in the corrugated region surrounding the directly excited area. Here the intensity shows a radially decreasing profile as we move away from the grating center. Such a distributed fluorescence is radiated both in air and in the substrate thanks to a mechanism of diffraction of BSW-coupled emission. Such image provides evidence that BSW-CE propagates along the surface before being diffracted, thus resulting in delocalization of the collected fluorescence far away from the original location of the excited emitters. This effect has also been found for surface plasmons over a smaller length scale, and critically considered in SPP-mediated imaging [9, 10].

The BFP of the oil immersion objective provides more details about the beaming effect by revealing the angular distribution of fluorescence. Figure 3.8 shows the fluorescence images of BFP in three different conditions: the BFP obtained by exciting fluorescence on a flat multilayer (a) reveals the band structure of the multilayer structure showing an homogeneous inner circle corresponding to the directly transmitted light surrounded by a system of circles beyond the light line corresponding to the TE and TM modes sustained by the 1DPC and leaking into the substrate. Figure 3.8b and c are obtained when the fluorescence is excited within the inner spacer of a concentric ring grating fabricated on top of a 1DPC with 16 layers (b) and 22 layers (c). In the latter case the leakage radiation associated to the BSW mode is almost completely suppressed.

The bright spot visible in the center of the BFP ($k_x/k_0 = 0$, $k_y/k_0 = 0$) corresponds to fluorescence that is beamed out normally with respect to the surface. The beaming effect can be explained as follows: as already observed in literature related to SPPs, [11, 12] part of the energy emitted by the excited sources (characterized by randomly oriented dipolar momenta) is transferred to BSW modes that radially propagates far away from the illuminated area along all the directions parallel to the truncation interface. The corrugation surrounding the emitter scatters the BSW-coupled radiation out of the surface and due to the specific periodicity of the corrugation the beaming effect occurs as consequence of the constructive interference along a specific directions. BSW can be therefore considered as near-field energy carriers that, thanks to the well defined momentum, can act as an efficient tool to redirect light in the desired direction. As concerns the diffraction efficiency, thanks to the FEM model we found that a relief grating fabricated with a dielectric material whose refractive index is higher than that of SiO_2 can extract the BSW-CE more efficiently. We therefore fabricated the grating by Electron Beam Lithography on a negative tone photoresist (MaN 2401 from MicroChem) spinned at 6000 rpm for 30 s (thickness 80 nm, $n_r = 1.6$). The experimental observation of the beam produced by the polymeric grating is reported in Fig. 3.9 and shows a significant improve of the extraction efficiency.

Fig. 3.9 Back focal plane image showing the fluorescence collected upon local illumination of the flat inner spacer of the polymeric grating. The dielectric loading mechanism due to the deposition of the thin polymeric layer produces a redshift for the BSW dispersion such that a wider wavelength range of the dye emission spectrum can be BSW-coupled and then beamed. As a result, a more intense fluorescence beaming can be obtained over a larger spectral range

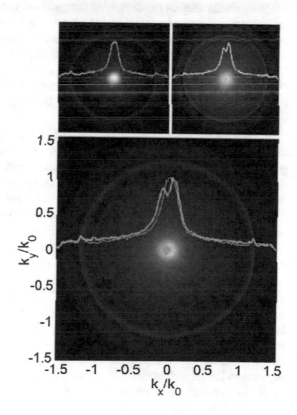

As already discussed the beaming effect can be explained as superposition of all the diffracted components of BSW-CE radially propagating from the center. As a result, the effect can be observed only when the fluorescence is excited in the center of symmetry of the structure. The mechanism can be explained by means of a simple geometrical model assuming that the circular grating is composed by a multiple set of linear gratings radially oriented [5]. We can therefore apply the Bragg's law along each azimuthal direction and considering the BSW coupled fluorescence as the incident radiation. As observed in the previous chapter, a linear grating generates a circular branch (+1 diffraction order) obtained by shifting the ring corresponding to the BSW-CE by an amount equal to the grating vector module [13]. By drawing the circular branches generated along all the azimuthal directions, we observe that a circular caustic appears in the Fourier plane, as a result of the superposition in a given locus of all the diffracted branches.

The diameter of the resulting circular caustic depends on the grating vector $G = 2\pi/\Lambda_g$. The minimum of the diameter is obtained when G equals the BSW-CE wavevector and the circular caustic collapses in a point in the center, corresponding to a bidimensionally collimated beam with very low divergence. In order to test the geometrical model we fabricated several circular gratings with different periods Λ_g. For each grating momentum we calculated analytically the corresponding Fourier pattern. The results are shown in Fig. 3.10.

The directionality of the diffracted fluorescence can be indirectly confirmed also by looking at the fluorescence image of the 1DPC surface when a low numerical aperture objective is employed. Figure 3.11 is obtained by exciting fluorescence with a collimated laser beam illuminating illuminating homogeneously the field of view. Fluorescence is collected by a $NA = 0.2$ objective operating in air. The grating with the smaller period ($\Lambda_g = 460$ nm, indicated with a white arrow) generates a caustic ring larger than the numerical aperture of the collection objective and therefore appears darker than the other rings.

If we look in detail at what happens when we excite fluorescence only in the center of the inner spacer, we observe that fluorescence is radiated from the whole patterned area surrounding the directly excited central spot (Fig. 3.12a). In order to collect only diffracted fluorescence and prevent leakage radiation ot contribute to the image formation we employed a low NA objective operating in air. We can analyze the intensity profile taken across the yellow dashed line in Fig. 3.12a. The profile (Fig. 3.12b) shows a central peak within the inner spacer corresponding to the directly excited spot that contains fluorescence radiated with no coupling with the surface modes. Surrounding the central peak are visible two side lobes whose intensity decreases as we move far from the center. The lobes are due to the BSW-CE that is diffracted by the corrugation surrounding the inner spacer. The intensity of the lobes drops down to zero outside the patterned area, where no diffraction occurs. The fluorescence pattern is circularly symmetric with respect to the center of the structure, reflecting the random orientation of the dipoles excited by the laser spot. Nevertheless, the BSW-CE diffracted should maintain the polarization state of the BSW mode, that depends on the azimuthal direction of propagation. We can observe such dependence by placing a polarization analyzer along the collection path. Figure 3.12c–e confirm

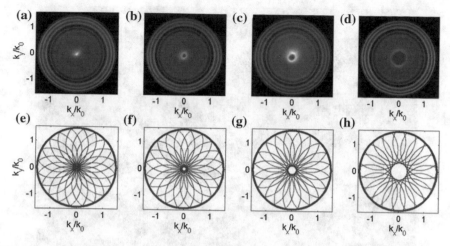

Fig. 3.10 Different BFP images of circular gratings on *regular* multilayers are compared to corresponding schematic BFP patterns calculated by a simple geometrical model. All measured BFPs are obtained by locally exciting multiple fluorescent emitters in the center of several circular gratings ($D = 5\,\mu m$) with different periods Λ_g. The corresponding calculated patterns are obtained by applying the Bragg law to the BSW-coupled fluorescence ring as diffracted by multiple wedges of the circular grating. For each wedge, a circular branch associated to the first order diffracted BSW is drawn. The superposition of all those diffraction branches produces a caustic circle whereon all the diffracted intensities sum up. The diameter of the caustic circle depends on the grating period Λ_g and indicates the *angular shape* of the diffracted fluorescence out of the multilayer surface. When the caustic circle collapses into a single point, the BSW-coupled fluorescence is almost completely beamed normally to the sample surface, with low divergence. **a, e** grating period $\Lambda_g = 560$ nm. **b, f** grating period $\Lambda_g = 520$ nm. **c, g** grating period $\Lambda_g = 500$ nm. **d, h** grating period $\Lambda_g = 460$ nm. Copyright Nature Publishing Group. Reprinted from [5]

that the diffracted light is TE polarized (the blue arrows indicate the orientation of the polarization analyzer). This result further confirms the BSW-assisted characteristics of the beamed fluorescence. The mechanism of BSW-assisted radiation extraction here described is somehow complementary to the BSW focusing mechanism assisted by the circular grating.

3.2.1 Experimental Quantification of the Extraction Enhancement Factor

Results shown above suggest that the collection of fluorescence at low NA would be significantly improved thanks to the BSW-assisted beaming effect. In this section we attempt a quantitative estimation of the fluorescence collection improvement.

In order to quantify the enhancement in fluorescence collection, we setup an original scanning microscope as illustrated in Fig. 3.13 and described in [5]. This setup is aimed at quantifying the amount of light radiated from a specific location

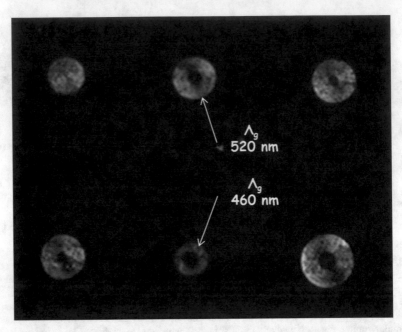

Fig. 3.11 Wide field fluorescence image of the surface of the 1DPC patterned with six circular gratings having different periods the same periods discussed in Fig. 3.10. The white arrows indicate two gratings with periods $\Lambda_g = 520$ and 460 nm respectively. The different fluorescence intensity is due to the different output angle of beamed fluorescence. The second one has indeed an output angle out of the NA of the collection objective

of the sample and directed along a specific direction, and eventually analyze its spectral composition. To perform such analysis, the microscope has to provide at the same time both spatial, angular and spectral resolution. The spatial resolution is obtained by focusing the laser beam ($\lambda = 532$ nm) into a diffraction limited spot by mean of an oil immersion objective (NA = 1.49), in such a way that fluorescence is locally excited. The fluorescence is collected by means of the same objective and spectrally filtered by an edge filter (RazorEdgeH Longpass 532) to filter out the laser radiation. A lens placed along the collection path produces a magnified image of the BFP of the objective onto a plane (here called BFP) B where a multi-mode fiber end is located. The fiber can be positioned in any position of the BFP image by means of a two dimensional translational stage, so that we can select the specific portion of BFP to be injected into the fiber. Since the BFP image is proportional to the angular distribution of the light collected, selecting a specific region of the BFP corresponds to select a specific direction of propagation of light, thus providing an angular resolution that depends on the relation between the fiber inlet size and the magnification of the BFP. As expected, the higher the angular resolution the lower the intensity detectable, so that a trade-off is needed. In our configuration we found an acceptable fluorescent signal when the inlet of the fiber superposes to a BFP region slightly smaller than NA = 0.1. In the specific, we located the fiber end at the center

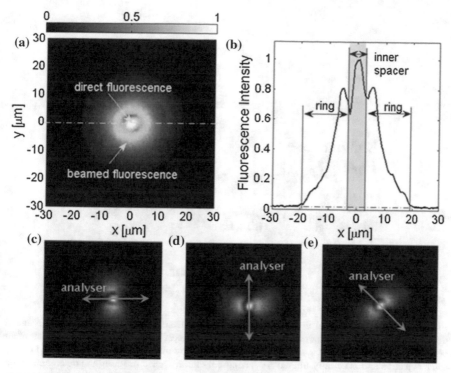

Fig. 3.12 **a** Direct image of fluorescence excited by a laser beam focused onto the flat inner spacer of a circular grating ($D = 5\,\mu$ m, $\Lambda_g = 520\,$nm) fabricated on *low leakage* multilayer. **b** cross-sectional fluorescence intensity along the horizontal dashed line in (**a**). **c–e** same as in (**a**) with the addition of a polarization analyzer in collection, whose orientation is sketched by the *arrows* in the figures. Copyright Nature Publishing Group. Reprinted from [5]

of the BFP image, so that normally propagating radiation is injected. The fiber is then connected to a spectrometer (Acton SpectraPro 300, Princeton Instruments).

The sample can be scanned with respect to the laser spot along the x and y directions by means of piezo-actuated translators. At each scan position the fluorescence excited and collected by the fiber is measured by the CCD of the spectrometer and the intensity spectrum is stored, so that each point of the resulting image contains the entire spectrum of the radiation propagating along the desired direction. Figure 3.14 shows an exemplary image obtained with the above described setup. It consists of a matrix of 150×150 pixels, with an integration time of 5 ms for each pixel. The laser illumination is kept at low power to avoid rapid photo-bleaching of the emission during the scan. The scanned region contains a circular grating structure and the false colors correspond to the intensity integrated over a spectral range from $\lambda = 570$ nm to $\lambda = 595$ nm, after background subtraction. As clearly observable, when the laser spot is located at the center of the concentric ring grating a significant increase of fluorescence intensity is observed, and the grating structure is distinguishable from the flat area because of increased scattering. The image clearly shows that the BSW-

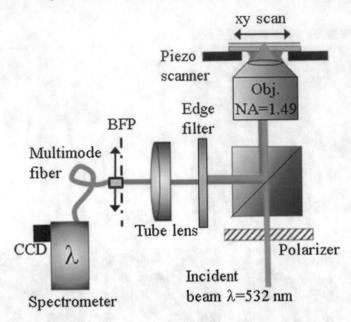

Fig. 3.13 Schematic view of the BFP scanning fluorescence microscope. The sample is scanned through a focused laser beam that locally excites fluorescence. Illumination is done by means of an oil immersion objective (NA = 1.49). The radiation leaking into the substrate is collected by the same objective and spectrally filtered for having only fluorescence reaching the detector. A tube lens produces a magnified BFP image over a remote plane where a collection fiber (core diameter 50 μm) can be accurately positioned. The fiber is movable along two directions, in such a way that light can be fiber collected from selected regions of the BFP image. Collected light is sent to a spectrometer equipped with a CCD camera. Copyright Nature Publishing Group. Reprinted from [5]

assisted beaming efficiently enhances the fluorescence intensity collected at NA < 0.1 only emitters located at the center of the inner spacer are excited.

Figure 3.14b shows a comparison between the fluorescence intensity spectra integrated over two regions of the image in Fig. 3.14a marked by a green square (flat region, green spectrum) and a black square (inner spacer, black spectrum). Since the beaming effect appears to amplify more a certain spectral region, several spectral ranges are considered, as sketched in the Fig. 3.14b. For each spectral range (identified by a colored vertical bar), the enhancement factor $I_{spacer}/I_{outside}$ is calculated, wherein I_{spacer} and $I_{outside}$ refer to the fluorescence intensities integrated over the corresponding spectral range and over spatial regions defined by the black box and the green box in Fig. 3.14a, respectively. A set of values for the ratio $I_{spacer}/I_{outside}$ is obtained by making the green box sequentially sampling the whole image, with the exclusion of the grating area.

Finally, the collected data set for each spectral range are represented as histograms, as shown in Fig. 3.14c–g. Each histogram provides information about the statistical distribution of the enhancement factor, clearly wavelength-dependent. For example,

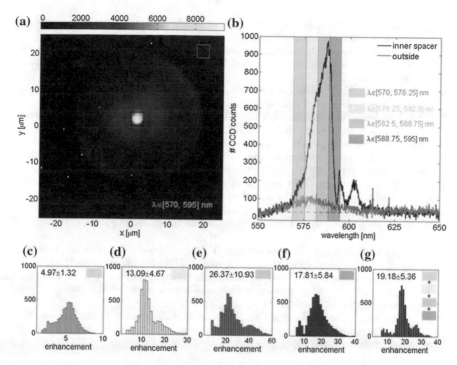

Fig. 3.14 **a** False-color image of the circular grating as obtained by means of the BFP scanning system. Intensity values are integrated over a spectral range from $\lambda = 570$ nm to $\lambda = 595$ nm. **b** Fluorescence spectra corresponding to the grating inner spacer and outside the grating. *Colored bars* indicate spectral intervals used for further analysis. **c–g** statistical distributions of the enhancement factor calculated over the corresponding spectral ranges. Average and standard deviation are indicated for each spectral range. Copyright Nature Publishing Group. Reprinted from [5]

the highest enhancement factor is estimated as 26.37 ± 10.93 and it is observed for wavelengths between $\lambda = 582.5$ nm and $\lambda = 588.75$ nm (Fig. 3.14e). The mean enhancement factor over the whole spectral range from $\lambda = 570$ nm to $\lambda = 595$ nm is estimated as 19.18 ± 5.36 (Fig. 3.14g).

References

1. J.-M. Yi, A. Cuche, E. Devaux, C. Genet, and T. W. Ebbesen, *Beaming Visible Light with a Plasmonic Aperture Antenna*, ACS Phot. 1, 365–370, (2014).
2. H. Aouani, O. Mahboub, E. Devaux, H. Rigneault, T. W. Ebbesen, J. Wenger, *Plasmonic antennas for directional sorting of fluorescence emission*, Nano Lett. 11 (6), 2400–2406, (2011).
3. H. Aouani, O. Mahboub, E. Devaux, H. Rigneault, T. W. Ebbesen, and J. Wenger, *Large molecular fluorescence enhancement by a nanoaperture with plasmonic corrugations*, Opt. Expr. 19(14), 13056–13062, (2011).

4. Y. C. Jun, K. C. Y. Huang and M. L. Brongersma *Plasmonic beaming and active control over fluorescent emission*, Nat. Comm. 2, 283, (2011).
5. A. Angelini, E. Barakat, P. Munzert, L. Boarino, N. De Leo, E. Enrico, F. Giorgis, H. P. Herzig, C. F. Pirri, E. Descrovi, *Focusing and Extraction of Light mediated by Bloch Surface Waves*, Sci. Rep. 4, 5428, (2014).
6. G.M. Lerman, A. Yanai and U. Levy, *Demonstration of nanofocusing by the use of plasmonic lens illuminated with radially polarized light* Nano Lett. 9, 2139–2143, (2009).
7. Z. Liu, J. M. Steele, W. Srituravanich, Y. Pikus, C. Sun and X. Zhang, *Focusing surface plasmons with a plasmonic lens*, Nano Lett. 5, 1726–1729, (2005).
8. A. Angelini, E. Enrico, N. De Leo, P. Munzert, L. Boarino, F. Michelotti, F. Giorgis and E. Descrovi, *Fluorescence direction assisted by Bloch surface waves on a one-dimensional photonic crystal*, New J. Phys. 15, 073002, (2013).
9. S. G. Aberra, J. Laverdant, C. Symonds, S. Vignoli and J. Bellessa *Spatial coherence properties of surface plasmon investigated by Young's slit experiment* Opt. Lett. 37, 2139–41, (2012).
10. S. G. Aberra, J. Laverdant, C. Symonds, S. Vignoli, F. Bessueille and J. Bellessa, *Influence of surface plasmon propagation on leakage radiation microscopy imaging* Appl. Phys. Lett. 101, 123106, (2012).
11. G. Rui, D. C. Abeysinghe, R. L. Nelson and Q. Zhan, *Demonstration of beam steering via dipole-coupled plasmonic spiral antenna.* Sci. Rep. 3, 2237 (2013).
12. H. Aouani, O. Mahboub, N. Bonod, E. Devaux, E. Popov, H. Rigneault, T. W. Ebbesen, J. Wenger, *Bright unidirectional emission of molecules in a nanoaperture with surface corrugations.* Nano Lett. 11, 637–644 (2011).
13. H. L. Offerhaus, B. van den Bergen, M. Escalante, F. B. Segerink, J. P. Korterik, and N. F. van Hulst, *Creating focused plasmons by noncollinear phase matching on functional gratings.*, Nano Lett. 5, 2144–2148, (2005).

Conclusions

The presented work shows that one dimensional photonic crystals can act as a suitable platform either for 2D optics or as energy mediators between far field radiation and localized emitters via near field coupling.

The low absorption of dielectric materials allows for designing structures able to sustain surface waves that can propagate for several hundreds of microns in the visible range. Along the path, BSW can be eventually manipulated by means of ultra-thin dielectric structures directly patterned on the surface of the 1DPC by standard lithographic techniques. Alternatively, freely propagating radiation can be converted into BSW (and *vice-versa*) by exploiting surface couplers that can eventually be focused in a subwavelength volume when a concentric ring grating is considered.

The influence of the photonic resonant structure on spontaneous emission has also been considered. Due to the high density of photonic states occurring at the truncation interface of the 1DPC, spontaneous emission is strongly modified. The resonant mode acts indeed as a preferential drain channel for electromagnetic energy, that is therefore radiated with the specific wave-vector given by the BSW dispersion relation. Along the propagation at the truncation interface, part of the energy leaks into the substrate preserving the transverse wave-vector component. The radiation pattern in far-field results in a highly collimated and dispersed beam.

Once the radiation is coupled with the BSW, it can be manipulated by means of surface structures, similarly to what happens with Surface Plasmon Coupled Emission.

An interesting chance is to convert the BSW-coupled fluorescence into a highly collimated beam that can be out-coupled along any desired direction by employing surface diffractive gratings and simply applying the Bragg's law. In particular, the case wherein a localized source is surrounded by a bull's eye antenna allows to maximize the directional extraction. The overall effect in far-field is that the spontaneous emission of a point-like source is converted from an almost isotropic pattern into a highly bi-dimensionally collimated beam.

The photonic structure discussed in this work can be an interesting resonant substrate for 2D light manipulation since offers some advantages with respect to typical photonic crystal structures and to metallic films. Moreover, due to the strong influ-

© Springer International Publishing AG 2017
A. Angelini, *Photon Management Assisted by Surface Waves on Photonic Crystals*, PoliTO Springer Series, DOI 10.1007/978-3-319-50134-5

ence over the spontaneous emission occurring close to the surface, it is a suitable platform for applications such as few molecules fluorescence sensing, efficient single photon extraction or, eventually, surface enhanced Raman scattering, since it can be easily integrated with plasmonic nano-antennas.

References

1. E. Descrovi, T. Sfez, L. Dominici, W. Nakagawa, F. Michelotti, F. Giorgis and H. P. Herzig, *Near-field imaging of Bloch surface waves on silicon nitride one-dimensional photonic crystals*, Opt. Expr. 16 (8), 5453–5464, (2008).
2. W. M. Robertson, G. Arjavalingam. R. D. Meade, K. D. Brommer, A. M. Rappe, and J. D. Joannopoulos, *Observation of surface photons on periodic dielectric arrays*, Opt. Lett. 18, 528–530 (1993).
3. D. Artigas and L. Torner, *Dyakonov Surface Waves in Photonic Metamaterials*, Phys. Rev. Lett. 94, 013901, (2005).
4. F. Michelotti, A. Sinibaldi, P. Munzert, N. Danz and E. Descrovi, *Probing losses of dielectric multilayers by means of Bloch surface waves*, Opt. Lett. 38, 616–618, 2013.
5. C. Zhao and J. Zhang, *Plasmonic Demultiplexer and Guiding*, ACS Nano 4, 6433 (2010).
6. M. S. Kim, T. Scharf, C. Etrich, C. Rockstuhl, and H. P. Herzig, *Longitudinal-differential interferometry: direct imaging of axial superluminal phase propagation.*, Opt. Lett. 37 (3), 305–307 (2012).
7. K. G. Lee, X. W. Chen, H. Eghlidi, P. Kukura, R. Lettow, A. Renn, V. Sandoghdar and S. Gotzinger, *A planar dielectric antenna for directional single-photon emission and near-unity collection efficiency.*, Nat. Photon. 5, 166–169, (2011).
8. C. J. Regan, R. Rodriguez, S. C. Gourshetty, L. Grave de Peralta, and A. A. Bernussi, *Imaging nanoscale features with plasmon-coupled leakage radiation far-field superlenses*, Opt. Expr. 20 (19), 20827–20834, (2014).
9. M. F. Templin, D. Stoll, M. Schrenk, P.C Traub, C.F. Vöhringer and T.O. Joos, *Protein microarray technology*, Trend Biotechnol 20, 160, (2002).
10. K. Iida and I. Nishimura, *Gene expression profiling by DNA microarray technology*, Crit. Rev. Oral Biol. Med. 13, 35, (2002).
11. C.R. Taitt, G.P. Anderson and F. S. Ligler, *Evanescent wave fluorescence biosensors*, Biosens. Bioelectron. 20, 2470, (2005).
12. J.R. Lakowicz, *Principles of Fluorescence Spectroscopy Third Edition*, Springer, (2006).
13. L. A. Tessler and R.D. Mitra,*Sensitive single molecule protein quantification and protein complex detection in a microarray format*, Proteomics 11, 4 (2011).
14. S. George, V. Chaudhery, M. Lu, M. Takagi, N. Amro, A. Pokhriyal, Y. Tan, P. Ferreira and B. T. Cunningham, *Sensitive detection of protein and miRNA cancer biomarkers using silicon-based photonic crystals and a resonance coupling laser scanning platform.* Lab Chip 13, 4053, (2013).

© Springer International Publishing AG 2017
A. Angelini, *Photon Management Assisted by Surface Waves on Photonic Crystals*, PoliTO Springer Series, DOI 10.1007/978-3-319-50134-5

15. L. Han, D. Zhang, Y. Chen, R. Wang, L. Zhu, P. Wang, H. Ming, R. Badugu and J. R. Lakowicz, *Polymer-loaded propagating modes on a one-dimensional photonic crystal.*, Appl. Phys. Lett. 104, 061115 (2014).

16. S. Kim, H. Kim, Y. Lim, B. Lee, *Off-axis directional beaming of optical field diffracted by a single subwavelength metal slit with asymmetric dielectric surface gratings*, Appl. Phys. Lett. 90, 051113, (2007).

17. H. J. Lezec, A. Degiron, E. Deveaux, R. A. Linke, L. Martin-Moreno, F. J. Garcia-Vidal, T. W. Ebbesen, *Beaming light from a subwavelength aperture*, Science 297, 820–822, (2002).

18. R. Lopez-Boada, C. Regan, D. Dominguez, A. Bernussi, L. Grave de Peralta, *Fundaments of optical far-field subwavelength resolution based on illumination with surface waves*, Opt. Expr. 21, 11928–11942, (2013).

Printed in the United States
By Bookmasters